"Cases on Global IT Applications and Management: Successes and Pitfalls

Felix B. Tan///
University of Auckland, New Zealand

 **Idea Group Publishing**

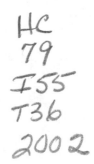 **Information Science Publishing**

Hershey • London • Melbourne • Singapore • Beijing

Acquisition Editor:	Mehdi Khosrowpour
Managing Editor:	Jan Travers
Development Editor:	Michele Rossi
Copy Editor:	Maria Boyer
Typesetter:	LeAnn Whitcomb
Cover Design:	Deb Andre
Printed at:	Integrated Book Technology

Published in the United States of America by
 Idea Group Publishing
 1331 E. Chocolate Avenue
 Hershey PA 17033-1117
 Tel: 717-533-8845
 Fax: 717-533-8661
 E-mail: cust@idea-group.com
 Web site: http://www.idea-group.com

and in the United Kingdom by
 Idea Group Publishing
 3 Henrietta Street
 Covent Garden
 London WC2E 8LU
 Tel: 44 20 7240 0856
 Fax: 44 20 7379 3313
 Web site: http://www.eurospan.co.uk

Library of Congress Cataloging-in-Publication Data

Tan, Felix B., 1959-
   Cases on global IT applications and management : successes and pitfalls / Felix Tan.
      p. cm.
   Includes bibliographical references and index.
    ISBN 1-930708-16-5 (cloth)
      1. Information technology--Government policy. 2. Information
technology--Management. I. Title.

HC79.I55 T36 2001
004--dc21   2001039456

British Cataloguing in Publication Data
A Cataloguing in Publication record for this book is available from the British Library.

Cases on Global IT Applications and Management: Successes and Pitfalls

Table of Contents

Preface

> *"Globalization is not an objective but an imperative, as markets and geographical barriers become increasingly blurred and even irrelevant"*
>
> John Welch Jr.
> CEO
> General Electric

These words by John Welch Jr., CEO of General Electric, were said at the beginning of the 1990s–the decade in which we witnessed the exponential growth of the Internet and e-business. Globalization took on a new meaning with success stories like Amazon.com and eBay. Even small companies in geographically disadvantaged countries have been able to penetrate international markets as the Internet creates a level playing field. For example, jenniferann.com—a small company that sells New Zealand-made lingerie and accessories—operates out of Pokeno, a small farming township south of Auckland, the country's major city. These successes are not restricted to just the dot-coms. Many companies have also effectively organized their global operations using the Internet and other information technologies (ITs). An example of such a company is United Parcel Service (UPS) its global operations are a classic case study on how the Internet and IT in general can be used to support and shape the business.

As organizations continue to compete globally in the new millennium, the effective deployment and exploitation of information technology will create the difference between those that are successful and those that are not. What lessons are there to be learned from organizations that run global IT operations and deploy IT in support of their global business operations? What lessons are there to be learned in the development, implementation, use and management of IT from public and private organizations operating in countries other than their own?

This book brings together 14 cases that report on these aspects of global IT applications and management. These cases are based on the actual experiences of business and IS professionals in large and small organizations. The primary mission of this book is to document successes and pitfalls of IT applications and management in organizations that compete in the global arena, and in public and private organizations around the globe. Cases included in this publication can be valuable teaching and research resources for educators and researchers in the global IT field. The cases are organized into three sections. The following paragraphs provide summaries of the cases included in each of the three sections.

Section I: Global Teaching Cases

The Australasian Food Products Co-Op: A Global Information Systems Endeavour

International information systems (IIS) have taken on increased importance as organizations develop and refine their global operations. This case is based on the history of an IIS project in a real transnational enterprise. It describes an "information system migration" following the development of a Global Business Strategy of a Multi-National enterprise through various stages. The Co-op, a very large organization, had to grapple with sufficient command and control to mount synchronised international marketing and logistics operations in an increasingly competitive global market. At the same time, management in its international sites demands local autonomy. The case describes the Co-op's attempt at developing an international information system that provides a balance between central control and local flexibility. It focuses on the management issues and difficulties faced by the organization in delivering such a global system. Failure of the IIS to adapt to the organization's strategy changes sets up a field of antagonistic forces, in which business resistance eventually defeated all attempts by the information technology people to install a standard global information system.

LXS Ltd. Meets Tight System Development Deadlines via the St. Lucia Connection

This case describes how a LXS Ltd., a Toronto-based software house, meets a tight systems development deadline for their new product called Estitherm, a Web-based software tool that supports heat loss calculations for architectural engineers designing structures. Estitherm's development requires sophisticated Java programming skills. The project stalls when LXS is unable to hire enough additional programmers to be able to meet the development deadlines dictated by competition. Through lucky coincidence, LXS' Chief Scientist stumbles onto a pool of Java talent while vacationing on the Caribbean island of St. Lucia. Negotiations follow, a contract is signed and the project is quickly brought to successful completion with the aid of Caribbean programmers, working via the Internet. Similar contract arrangements hold the promise for improved economic conditions in Caribbean nations and can reduce software backlogs for companies in developed nations, but better mechanisms are needed to bring together buyers and sellers of IT services.

Success and Failure in Building Electronic Infrastructures in the Air Cargo Industry: A Comparison of The Netherlands and Hong Kong SAR
This case describes the genesis and evolution of two IOSs in the air cargo community and provides information that lets students analyze what led one to be a success and one to be a failure. The two examples are from the Netherlands and Hong Kong SAR. The case emphasizes the complex, institutional and technical choices by the initiators of the system in terms of their competitive implications that were the main causes for the systems' fate. The case thus argues that it was the institutional factors involved in the relationships of the stakeholders that led to the opposite manifestations of the two initiatives and not the available technology nor a lack of talent in producing sufficiently good systems. The case therefore lends itself to advocate that also non-technological factors should be taken into account when designing and implementing interorganizational information systems.

Ford Mondeo: A Model T World Car?
This case weighs the advantages and disadvantages of going global. Ford presented its 1993 Mondeo model, sold as Mystique and Contour in North America, as a 'world car.' It tried to build a single model for all markets globally to optimize scale of production. This required strong involvement from suppliers and heavy usage of new information technology. The case discusses the difficulties that needed to be overcome as well as the gains that Ford expected from the project. New technology allowed Ford to overcome most of the difficulties it had faced in earlier attempts to produce a world car. However, IT was flanked by major organization changes within Ford. Globalization did not spell obvious success though. While Ford may in the end have succeeded in building an almost global car, it did not necessarily build a car that was competitive in various markets. The Mondeo project resulted in an overhaul of the entire organization under the header of Ford 2000. In terms of Ford's own

history, the Mondeo experience may not be called a new Model T, but does represent an important step in Ford's transformation as a global firm.

Lone River Winery Company: A Case of Virtual Organization and Electronic Business Strategies in Small and Medium-Sized Firms
Electronic networks and virtual organizing capabilities are shaping the competitive performance of small firms in the global information economy. As hardware becomes increasingly affordable, soft assets and strategies will determine the real winners. Successful small and medium-sized firms will be those with distinctive skills to manage the unique features of both the electronic marketplace and the enabling infrastructure. This chapter discusses a case example of Internet infrastructure and e-business strategy management in small and medium-sized firms (SMFs). The case focuses on the key features of electronic business strategy using a virtual organizing framework. Based in the Swan Valley region of Perth, W. Australia, Lone River Winery Co. Ltd., has over the past five years employed the Internet to extend its business scope beyond the Australian wine market.

Dancing with a Dragon: Snags in International Cooperation Between Two IT Companies
International strategic alliances are an increasingly popular way for companies to expand their operations beyond national boundaries. For small and medium-sized enterprises, this route provides interesting characteristics that match their resources and ambitions. It is particularly appropriate to small companies in the information and communication technologies sector. The short product life cycles, rapid market developments and increasing product complexity enhance the attractiveness of international alliances. However, success does not come automatically in international strategic alliances. This case applies the aspects of strategic fit, resource fit, cultural fit and organizational fit to analyze the cooperation between a small Dutch IT company and a major Chinese Internet contents provider. Their cooperation does not run smoothly. The chapter identifies several misfits in this case study.

Section II: Regional Teaching Cases

Success in Business-to-Business E-Commerce: Cisco New Zealand's Experience
The growth of business-to-business e-commerce has highlighted the importance of computer and communications technologies and trading partner trust for the

development and maintenance of business relationships. The case study of Cisco New Zealand permits students to learn about: (i) the factors that influence the successful trading partner relationships in business-to-business e-commerce participation; (ii) different types of trading partner trust relationships and their impact on the benefits of business-to-business e-commerce applications; (iii) the implementation of Cisco Connection Online (CCO) and, in particular, what made Cisco's implementation successful and the challenges (issues) they face; (iv) the different roles of trading partner relationships and their impact on trust, the different forms of trust experienced in Cisco New Zealand and Compaq New Zealand using CCO, and (v) successful management practices in the context of CCO and their impact on inter-organizational trust in business-to-business e-commerce participation.

Geographical Information System (GIS) Implementation Success: Some Lessons from the British Food Retailers

Geographical Information Systems (GISs) are becoming more prevalent for retailers in their use for both operational day-to-day and strategic long-term decision-making. This chapter presents the results of in-depth case studies, reflecting upon the GIS implementation experiences of key UK food retailers. Two retailers were studied that have different positions in the marketplace and have employed GIS to varying degrees. The theory developed shows the factors that are significant in the implementation of a GIS in retailing organizations. It reflects upon the experiences of market leaders and the retailers that struggle to keep up with them.

Implementing and Managing a Large-Scale E-Service: A Case on the Mandatory Provident Fund Scheme in Hong Kong

This case concerns a recently launched retirement protection scheme, the Mandatory Provident Fund (MFP), in Hong Kong. Service providers, employers, employees and the government are the four main parties involved in MPF. The service has been implemented in two versions, a bricks model and a clicks model. The former is based on a conventional model with paper-based transactions and face-to-face meeting. The focus of this case, however, is on latter, which introduces MPF as a service in an e-environment that connects all parties electronically and conducts all transactions via the Internet or other computer networks. The case discusses the MPF e-business model and its implementation. The chapter analyzes the differences between the old and the new model; and highlights the chief characteristics and benefits of the e-business model as they arise from the emerging digital economy. It also discusses some major problems, from both managerial and technical perspectives, that have occurred during the phases of implementing and launching the new service.

Enabling Electronic Medicine at Kiwicare: The Case of Video Conferencing Adoption for Psychiatry in New Zealand

This case study highlights the factors influencing adoption of telemedicine utilizing video conferencing technology (TMVC) within a New Zealand hospital known as KiwiCare. The author applied the technological innovation literature to provide insights into TMVC adoption within KiwiCare. The case study indicates that the weak presence of critical assessment into technological innovation factors prior to the adoption decision can lead to its weak utilization. Factors like complexity, compatibility and trialability were not assessed extensively by KiwiCare and would have hindered TMVC adoption. TMVC was mainly assessed according to its relative advantage and to its cost effectiveness along with other facilitating and accelerating factors. This is essential but should be alongside technological and other influencing factors highlighted in the technology innovation literature.

The Role of Virtual Organizations in Post-Graduate Education in Egypt: The Case of the Regional IT Institute

The role of virtual organizations and virtual teams are rapidly spreading worldwide in the related aspects to learning and human resources development. This has led to the establishment of a large number of regional and global learning consortiums and networks aiming to provide quality knowledge and information dissemination vehicles to an ever-growing community of seekers that is online, active and eager to increasingly learn more. This case reports on the experience of Egypt's Regional IT Institute in the field of education and training. The Regional IT Institute was established in 1992 targeting the formulation of partnerships and strategic alliances to jointly deliver degree (academic) and non-degree (executive) programs for the local community, capitalizing on the enabled processes and techniques of virtual organizations. The case provides many lessons to be replicated that demonstrate the opportunity to expand in exchanging the wealth of knowledge across societies using a hybrid of forms for virtual organizations and virtual teams.

Section III: Research Cases

IT Industry Success in Small Countries: The Cases of Finland and New Zealand

Given the importance of the information technology industry in today's global economy, much recent research has focused on the relative success of small countries in fostering IT industries. This chapter examines the factors of IT industry success in small developed countries, and compares two such countries, Finland and New Zealand. Finland and New Zealand are alike in many respects, yet Finland's IT industry is more successful than New Zealand's. Three major factors that impact on the development of a successful IT industry are identified: the extent of government IT promotion, the level of research and development, and the existence of an education system that produces IT-literate graduates.

Identifying Supply Chain Management and E-Commerce Opportunities at PaperCo Australia

Established supply chain management techniques such as Quick Response (QR) or Customer Relationship Management (CRM) have proven the potential benefits of re-organizing an organization's processes to take advantage of the characteristics of electronic information exchange. This case presents the experiences of a large Australian paper products manufacturer in implementing an electronic document exchange strategy for supply chain management. It highlights the drivers for change that spurred their actions, and describes the issues associated with trying to support existing and future requirements for document exchange across a wide variety of trading partners. The experiences of PaperCo will be relevant to organizations with diverse trading partners, especially small to medium enterprises (SMEs).

Barriers to IT Industry Success: A Socio-Technical Case Study of Bangladesh

In Bangladesh, information technology (IT) use is still in a backward stage in terms of information generation, utilization and applications. A dependable information system has not been developed for the management and operation of thegGovernment machinery and large volume of data transactions in the public/private sector organizations. There is a lack of locally and externally generated information needed for the efficient performance of the government, production, trading, service, education, scientific research and other activities of the society. This case study of the IT scenario in Bangladesh discusses the challenges, analyzes the key issues that may be barriers to the success of its IT industry. It discusses the inherent strengths that can be used as the launching pad for making Bangladesh a potential offshore source of software and data processing services.

The Case Method and Teaching Case Note

Although there are numerous definitions, Yin (1994) defines a case study as:
> "… an empirical inquiry that investigates a contemporary phenomenon within its real-life context, especially when the boundaries between phenomenon and context are not clearly evident." (p. 13)

The case study method is particularly well suited to investigating information systems in organizations (Myers, 1997) and is the most common qualitative method used the field (Orlikowski and Baroudi, 1991; Alavi and Carlson, 1992).

At the same time, the case method is also a popular pedagogical approach to teaching information systems to students. It engages students and requires them to become active participants rather than passive observers. The real-life situations presented in cases encourage students to think independently and analyze the topic from all sides. The case method therefore heavily involves students in the learning process.

To support the teaching of the cases in this book, a teaching case note supplement is available to instructors. It highlights the learning objectives, provides an overview of the issues being studied and discusses the case setting. In addition, the teaching case notes supply suggestions as to how the class session may proceed as well as discussion questions. These cases can be used independently or be used with supplementary materials such as background literature as suggested in the teaching notes.

The Audience

We have witnessed a strong growth in the interest in global IT as a subject in the 1990s (Gallupe and Tan, 1999). This has led to the development of global IT workshops and courses. It is my hope that this casebook will be a valuable resource for such courses.

This book will suit students in advanced undergraduate and graduate programs in Information Systems. It can also be used in teaching aspects of International Economics or International Business, where information systems and technology are discussed. Cases in this publication can be an important resource in corporate management education programs. Whichever audience, the casebook can be used by itself, or it can serve as a supplement to global IT texts and readings.

There are a number of research cases in this book. The teaching cases have also been written based on the authors' field research. Research questions can be raised from these cases for further investigations.

Acknowledgments

The completion of an undertaking such as this would not have been possible without the contributions, support and assistance from a number of individuals. I would like to express my sincere gratitude to the contributors to the book. All of the authors of the cases included in this publication also served as referees for cases written by other authors. Thanks go to all those who provided constructive and comprehensive reviews. Special thanks also go to the publishing team at Idea Group Publishing. In particular to Jan Travers and Michele Rossi, who kept the project on schedule, and to Mehdi Khosrowpour, whose enthusiasm motivated me to initially accept his invitation to take on this project. In closing, I wish to thank all the authors for their insights and excellent contributions to this book as well as serving as reviewers. I also thank Mehdi Khosrowpour and Jan Travers at Idea Group for the ongoing professional support. Finally, I thank my wife and children for their love and encouragement throughout this project.

> **Felix B Tan, Ph.D.**
> **Auckland, New Zealand**
> **August 2001**

References

Alavi, M. and Carlson, P. (1992). A review of MIS research and disciplinary development. *Journal of Management Information Systems,* (8)4, pp. 45-62.

Gallupe, R.B and Tan, F.B. (1999). A Research Manifesto for Global Information Management. *Journal of Global Information Management,* (7)3, pp. 5-18.

Myers, M. D. (1997). Qualitative Research in Information Systems. *MIS Quarterly* (21)2, June, pp. 241-242. MISQ Discovery, archival version, http://www.misq.org/misqd961/isworld/. MISQ Discovery, updated version, September 12, 2000 http://www.auckland.ac.nz/msis/isworld/.

Orlikowski, W.J. and Baroudi, J.J. (1991). Studying information technology in organizations: Research approaches and assumptions. *Information Systems Research* (2), pp. 1-28.

Yin, R.K. (1994). *Case Study Research, Design And Methods*, 2nd Ed. Newbury Park: Sage Publications.

Section I
Global Teaching Cases

The Australasian Food Products Co-Op: A Global Information Systems Endeavour

Hans Lehmann
University of Auckland, New Zealand

EXECUTIVE SUMMARY

This chapter tells the case story of a Food Products Co-op from 'Australasia'[1] and their attempt to create a global information system. The Co-op is among the 20 largest food enterprises in the world, and international information systems (IIS) have taken on increasing importance as the organization expanded rapidly during the 1980s and even more so as the enterprise refined their global operations in the last decade. Set in the six years since 1995, the story demonstrates the many pitfalls in the process of evolving an IIS as it follows the Co-op's global business development. Two key findings stood out among the many lessons that can be drawn from the case: first, the notion of an "information system migration" following the development of the Global Business Strategy of the Multi-National enterprise through various stages; second, the failure of the IIS to adapt to the organization's strategy changes set up a field of antagonistic forces, in which business resistance summarily killed all attempts by the information technology department to install a standard global information system.

BACKGROUND

Marketing authorities for land-based industries (such as fruit growers, meat producers, dairy farmers, forestry, etc.) are often large companies with a strong international presence. The Australasian Food Products Co-op[2] (the 'Co-op') with some $5bn revenue is one of the largest. Like most of the others, the Co-op is a 'statutory monopoly,' as there is legislation that prohibits any other organization from trading their produce in international markets. With about a quarter of its revenue from raw materials and manufacturing outside Australasia, the Co-op is a mature transnational operator. Structured into nine regional holding companies, it has a presence in 135 offices in 40 countries. The 15,000 primary producers are organized into 35 cooperative 'Production Companies' (ProdCos), where they hold shares in proportion to their production. The ProdCos, in turn, own the Co-op. This tight vertical integration is seen as a big advantage. It allows the Co-op to act as one cohesive enterprise and to develop a critical mass needed in most of its major markets. Figure 1 shows this structure and the product flow.

The Co-op distributes its profit through the price it pays farmers (by way of the ProdCos) for their produce. Because of this, there is no 'profit' in the normal sense shown in the Co-op's accounts, which makes the traditional (financial) assessment of the Co-op's performance somewhat difficult. Similarly, because the Co-op lays down a demand forecast for each product by

Figure 1: Business Structure of the Co-Op

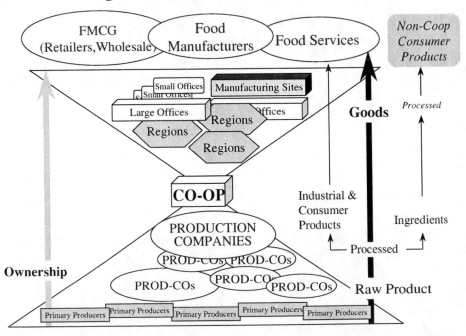

volume and time, which determines production schedule of the ProdCos, their performance is difficult to measure. In the absence of hard and fast facts, then, the interaction between the Co-op and its owners is a highly political one. Particularly the larger ProdCos have been pushing for some time to relax the Co-op's monopoly, so that they may export on their own account, arguing that they could achieve a higher return for their farmers. The Co-op counters this by pointing out that it has maximized the average return to all farmers: they could not allow a second party to "pick the eyes" of its most lucrative markets and leave the Co-op to deal with the difficult and marginal ones to the detriment of the majority of dairy farmers.

NATURE OF THE CO-OP'S BUSINESS

With the exception of 2% of its revenue, the Co-op deals exclusively with basic food products and their derivatives. It divides its operations into three main business segments. These are defined as follows:

1. *Consumer Market*—product is sold to the consumer in a consumer pack under a brand owned by the Co-op.
2. *Food Service Market*—product is supplied directly to third parties who prepare and generally serve food, meals, snacks away from home. This is a new and very fast-growing segment.
3. *Ingredients Market*—product is sold to third parties who are usually industrial food manufacturing companies; can also be any other industrial manufacturing companies (e.g., clear plastics in the North American market).

In addition, there is a large group of "other" markets, which is a mixture of non-brand sales, non-own-product sales and covers a very large variety of minority and speciality operations and markets. Figure 2 shows the proportions of the four business areas.

The Co-op's major markets are in Asia, the Americas and Europe,[3] which make up 85% of its revenue. Australasia,[4] the CIS,[5] and the Middle East[6] make up the rest. Figure 2 shows the distribution of the Co-op's business over the major regions.

The Co-op's Operations

There are, in principle,[7] two types of operations that the Co-op is involved in:

1. Distribution—this includes the sale efforts, the warehousing of product, as well as the logistics of obtaining and delivering it.
2. Manufacturing of product—mainly branded goods.

Figure 2: Mix of the Co-Op's Markets by Market/Industry Segment

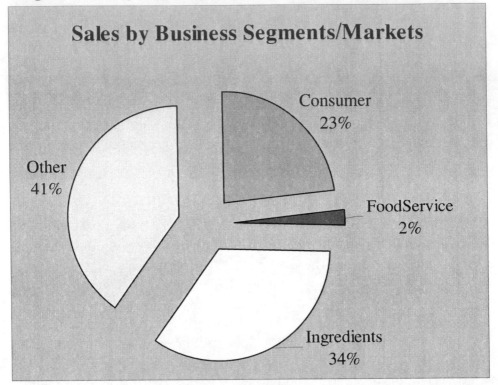

Figure 3: Geographical Distribution of the Co-Op's Business

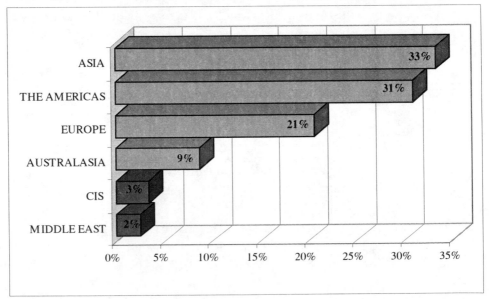

The manufacturing operation includes the manufacture from raw material as well as the re-packaging of product from mass transport units into end-consumer products. It does not, however, include activities that are carried out by the ProdCos, albeit to the Co-op's specifications:

(a) the primary manufacturing processes that results in the required variety of raw materials for further use;

(b) the manufacturing of end-product in the primary processing plants: although this occurs to the Co-op's specifications and orders, the actual process is outside the Co-op's management area.

Figure 4 illustrates the operations flow within the Co-op. Collecting the fresh product, all primary processing and some further processing (to produce end-product) takes place under the control of the Prod-Cos, on whom the Co-op places its orders. These are either for end-product (branded consumer products and ingredients) or raw material for further manufacturing in the regions. There, additional, local, raw materials may be used for the production of—mainly—branded consumer product. Customers in the three main markets (consumer, food-services and ingredients) place orders and take deliveries locally.

The main markets also have an influence on the Co-op's operation—the nature of the order-to-delivery cycle is different for each of them:

Figure 4: Schemata of the Co-Op's Operations

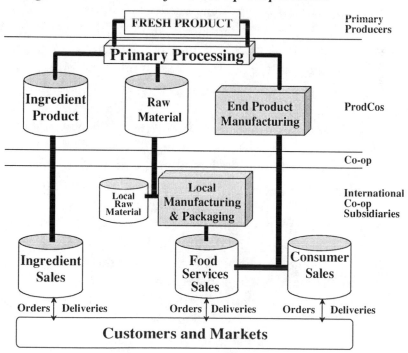

1. Consumer products are mostly sold through large retailers, and often special arrangements are made for each individual retailer; an example is the arrangement with Sainsbury's, a large food retailer in the UK: there the Co-op just stocks their shelving space within the Sainsbury's shop on a straight replenishment basis. Sainsbury's pays for it once it has passed through their EFTPOS system. Instead of a traditional cycle of order-delivery note-invoice-payment (and all the reconciliations in between), the Co-op's in-store stocks are reconciled against the (electronic) EFTPOS records. Payments are made electronically into the Co-op's account.
2. Food services are either:
    (a) individual enterprises that order by telephone or by collection; cash purchases are frequent; or they are
    (b) large franchises (fast foods, such as McDonalds, theme restaurants like Sizzlers) where often delivery is to central distribution points and orders are issued electronically or in the form of standing orders.
3. Ingredients are sold to industrial goods manufacturers; they are often a critical raw material and as such are required in a just-in-time fashion. Orders are often received electronically; and payment can be on delivery with corrective credit (if any) passed back at any time.

Two additional factors shaping the nature of the Co-op's operations are:
1. *Size of the Local Office*—large offices, dealing with large numbers of transactions tend to have different, more regimented systems than small ones, where flexibility and human interaction are often critical to business success, but are frequently inimical to formal systems;
2. *Environmental sophistication*—This has two, albeit interrelated aspects:
    (a) *Market sophistication*: Co-op revenue comes in equal parts from sophisticated markets like U.S. manufacturing, UK and Japanese food retailing on the one hand, and developing areas like Latin America and Asia where business processes are often somewhat simpler, but business practices are habitually very different and technologies can be a hybrid between ancient and leading edge. Figure 5 depicts the split in revenue between developing and developed countries and regions;
    (b) *Use of computerised operations support*: the Co-op operates across a variety of developmental stages, ranging from offices where more than four out of five staff work their own PC/terminal (as in the new office in the CIS, the former Soviet Union); the Latin America region, where less than one in six works his own PC/terminal; and to local offices like Egypt

Figure 5: Co-Op Revenue Between OECD and Developing Countries

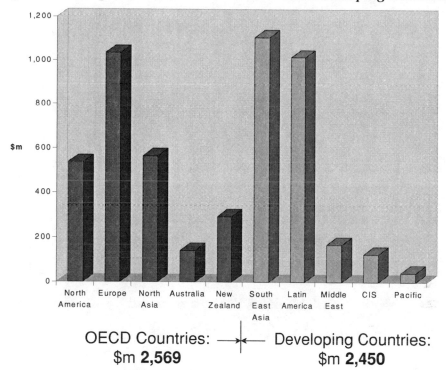

OECD Countries: ⎯⎯|⎯⎯ Developing Countries:
$m **2,569** $m **2,450**

(within the European, Mediterranean and North and West Africa regions) with virtually no computers at all involved in the operations.

The Co-op's business operations show a great variability with respect to a number of key variables. Revenue per employee varies widely from places that deal mainly in large volume, commodity-type products (as in the CIS) to places with high competition from a large domestic industry such as the UK and Europe. Not as wide is the variation of price per ton of product across the regions. However, it still oscillates from minus 30% through to plus 50% of the average price. Figure 6 accumulates all the deviation comparisons in one graph to convey the overall variability in the Co-op's business—as an indication of the large differences of the underlying regional operations.

External Environment

Over the last 20 years, the external environment for the Co-op's products and markets has become increasingly more demanding. Prior to that, the Co-op exported the vast majority of its produce to a small number of European countries who used to accept it all and in some instances even created customs barriers to protect this trade. Once the Common Agricultural Policy (CAP) was made binding for all EC members, however, all such barriers had to go

Figure 6: Variation of Key Indicators Across the Co-Op's Regional Operations (Deviation from Mean in % of the Mean)

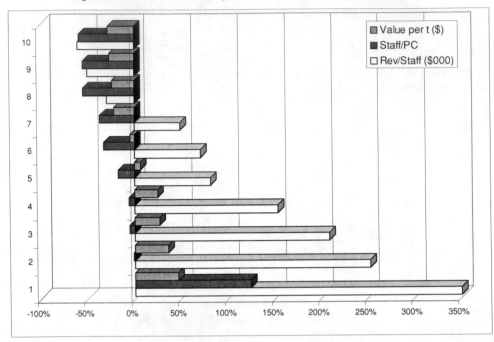

and other European countries flooded the market with heavily subsidized product. Virtually overnight the Co-op had to develop new markets. The food products industry in Australasia shrank dramatically in the initial years, but by 1990 had recovered equally dramatically. The remaining farmers are now very efficient high volume producers and the Co-op had turned itself from a producers club into a very effective international operator. Although still only a tenth the size of the largest international food corporations,[8] it has now moved into the same league as most major international food merchants.

The establishment of the World Trade Organization has brought an era of high opportunity for the Co-op. The new organization will for the first time introduce new trade rules for agricultural products, which will benefit the Co-op in two ways:

- They would dictate a gradual, but significant widening of the access to markets for all participants in the world trade; in the past there have been direct barriers (such as in the EC) and often indirect, non-tariff hurdles to market entry (such as the USA) which have very effectively hindered the Co-op's expansion.
- Limits have been introduced to the amount of export subsidies producer countries can grant their dairy industries.

This new set of rules works very much in favor of the Co-op with its high degree of vertical integration, its efficient producers and processing operations, and its well-functioning and widespread marketing organization.

BUSINESS BACKGROUND: THE CO-OP IN THE 1980s AND 1990s

Until the onset of the 1980s, the Co-op had maintained a small number of order-taking and warehousing operations mainly in the UK and Europe. With the dwindling away of those easy markets, the imperative was to find alternative places to sell dairy produce to. In rapid succession a number of subsidiary offices was set up and agencies nominated to deal with such markets as the U.S.

The main emphasis in this period was to shift product, to establish toeholds in new markets and to start building the foundations for a more permanent and long-term-oriented market presence later. The most effective way of achieving this was to staff these sales offices with capable and aggressive individuals and "let them get on with it." As long as they managed to "move" their quota, they had a high degree of freedom as to how they did it and with whom. This policy of local autonomy worked very well and achieved very much what was expected: within a decade the Co-op had built a presence in more than 30 countries and had managed to bring home, year after year, a regular, if not sensational, income for the all the Prod-Cos' primary producers and farmers.

At the onset of the 1990s, however, competition in the Co-op's main markets had become strong and increasingly global. With the emergence of global brands (such as Nestlé, Coca Cola, McDonalds), the Co-op needed to develop global brands themselves. For this, they had to have sufficient command and control to mount synchronized international marketing and logistics operations.

The main focus was on determining where to strike the balance between preserving the proven 'freedom' formula with the need to control and 'dictate' operations with a fast global reach. The high degree of local autonomy continues to maintain a great enthusiasm within local management, and the corresponding momentum still pushes the business at very good rates of expansion. On the other hand, global branding and parallel, multichannel distribution have proven their usefulness in launching new products into diverse markets with widely divergent levels of customer and/or retail sophistication such as in Europe and South East Asia.

The arrival of a new Chief Executive Officer in late 1994 began to instill purpose and urgency into this reorganization process. The Co-op began a concerted campaign to shift authority and control over branding and global marketing policy back to head-office. The CEO's vision was one of balanced central control and local flexibility. Figure 7 shows this development, using the Bartlett and Ghoshal[9] framework.

Part of this new outlook was a critical look at the Co-op's operations. In early 1995, partly for political reasons and partly to review the results of the past efforts to shift the strategic focus of the organization, the Co-op commissioned the Boston Consulting Group (BCG) to carry out an evaluation exercise of how the Co-op fares when compared with a benchmark of international best practice. Table 1 summarizes the results of their assessment report.

While overall very complementary of the Co-op's achievements, BCG were very critical of the amount of counterproductive politicking going on in the industry. They specifically focused on the efforts of the larger ProdCos to undermine the role of the Co-op and to move towards independence. Criticising this strongly, the consultants reiterated the importance of size and critical mass for an international food marketer.

Regarding its internal business functions, BCG assessed the Co-op itself a fit and effective enterprise. The Co-op was strongly commended on its strategic roles, but did not fare so well with regards to its organizational structure and, particularly, with respect to its management control and systems.

The Co-op accepted the critique of a fragmented, unclear and unfocused organization structure—an inheritance from its past rapid growth—and put in

Figure 7: The Co-Op's Migration of Global Business Strategy

Table 1: Summary of the BCG Evaluation of the Co-Op's
Management Practices

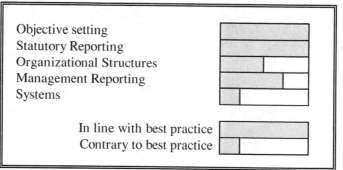

Objective setting
Statutory Reporting
Organizational Structures
Management Reporting
Systems

In line with best practice
Contrary to best practice

place a streamlined and efficient organizational structure. As a consequence of the low benchmark for management control systems, the CEO took a personal interest in the establishment of the *Food Information Systems and Technology (FIST)* project, which focused on business processes and control systems throughout the Co-op's operations worldwide.

THE IS LANDSCAPE IN 1995

During the 1970s and early 1980s, the Co-op had built up a sizeable IS department with a mainframe operation at the head office, linking up with all the main subsidiary offices and ProdCos throughout the country. Foreign activities were few and hardly needed computer support. The forced expansion drive in the late 1980s, however, led to an increased need by local operations to be supported with information systems. By the mid-1990s a number of regional offices had bought computers and software to suit their own, individual requirements (as shown in Figure 8).

Furthermore, by then each installation had selected their own software to suit their own requirements. However, reporting and ordering procedures back to the Co-op's head office were often manual and/or involved re-keying output from their local machines. Furthermore, different capabilities, varying data structures of the local systems, often meant that it was not always clear whether figures reported on a particular subject did indeed follow the same definition throughout all the countries. Monthly routine reports were received by the Co-op within anything from one week to three weeks after the event. Some special reporting requirements required widely differing efforts to implement and often could not be obtained across all subsidiaries.

The same difficulties (differing definitions) also plagued the attempt to install a brand/product reference system across some or all of the subsidiaries. Each local information system had implemented its own product numbering

Figure 8: The Co-Op's Configuration of Information Technology

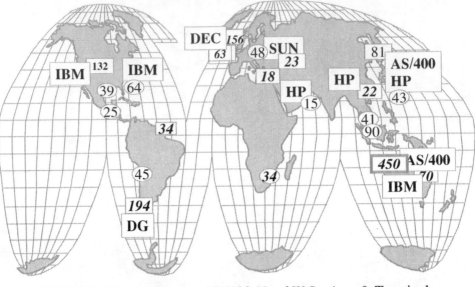

☐ MFR/Mini ◯ Lans *123*/123 Nr of *W*/*Stations* & Terminals

systems, based on its own, different logic, often in the form of composite numbers with an in-built structure.

Against this background of a proliferation of loosely, if at all, coordinated local systems and a declared will from the Co-op's centre to impose some more control over the enterprise as a whole, the IS Department in April 1995 established a 'Framework for Information Systems.' This was a declaration in principle that in the future there would be information technology standards throughout the Co-op's operating companies. In December 1995, this standards framework was extended to become the 'Charter for FIST,' intended to facilitate implementation of:

- Standard and common information definitions and formats across all the Co-op.
- Common information systems for the Co-op's business.
- Integrated flow of information from order to delivery, with electronic interfaces reducing the number of interrupt steps in this process.

The Co-op's Board formally ratified the FIST 'Charter' and the underlying framework. The Co-op's IS department subsequently interpreted this ratification as a mandate to develop common information systems for all of the Co-op's operating companies. The following tasks and subprojects were identified within FIST:

1. *Business Engineering*—this was to straighten out order processing and inventory control throughout the Group.

2. *The Global Applications Project (GAP)*—with the specific aim of designing and implementing common systems throughout the Group.
3. *Corporate Reporting*—i.e., common information standards for regular reports.
4. *Technology Standards Definition.*
5. *Tele-Communications Strategy.*

As a precursor, a 'Benchmarking Project' was carried out. This was a basic fact-finding exercise to define the current position and to establish a basic and common understanding of the Co-op's business operations and processes. The outcome of these exercises was a 'business model,' outlining where various business functions should be placed and the basis for this placement. A further outcome of the project was an estimate that the Co-op was spending approximately $80m on information systems, and it was expected that this would increase by not having common systems. The 'business model' was later named the *"Enquiry To Cash (ETC)"* model and would serve as the single process of supply and demand management across all of the Co-op's subsidiaries as well as the interface with the ProdCos. This is shown in Figure 9.

A first statement was also made at this stage about the style of the project, in response to a query by the head of the Latin America office - who was worried that the FIST project would, from HQ, dictate a new, common, global system which would make his considerable investment in information systems obsolete. The CEO assured him that "While the Group Information Systems project will initially focus on two pilot sites, all regions will be involved and kept informed." Furthermore, "each Region would be asked to nominate representatives who will be fully involved as members of the project team" and "the FIST project will concentrate more on information and common reporting standards" so that "there is as yet no intention of requiring that each region adopt the same computer system."

The first project strategy and plan for FIST foresaw the following main stages:
1. Development of a prototype system with a site which is reasonably representative for most of the Group's offices.
2. Implementation of the prototype in a small number of pilot sites; further adaptation of the prototype to make it functional as a global system.
3. Gradual 'roll-out' of the 'global system' into selected regions.

Estimated completion dates were late 1996, early 1997 and mid-1998 respectively.

North America, by the end of 1995, had started to embark on a review of its information systems. Their old IM system was becoming obsolete and the

Figure 9: The ETC Model of Business Operations

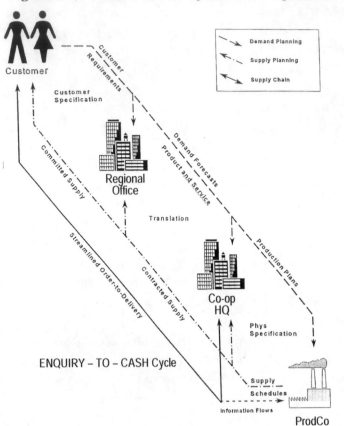

software was also in need of a functional upgrade. The South East Asia region was by then also looking to upgrade their fragmented PC-based installation with a more coherent information system to cope with the rapid growth the region was experiencing. Both sites thus became the natural candidates for the development of the prototype and also as pilot sites for further implementation:

- North America has a large ingredients market and would become the prototype for the Co-op offices serving this business sector.
- Singapore is the office looking after South East Asia; they have a large consumer market, which would make them a good prototype for all consumer and/or mixed business offices within the Group.

FIST began in earnest with the dispatch of a team to North America in March of 1996.

The North America Pilot

Having realized that their system needed upgrading, the North America Region was now very keen to go ahead with the replacement project as fast as possible. However, as this was to be the pilot for a global system, the FIST team was mainly interested in obtaining a thorough requirements specification for use as the foundation for further globalization.

Negotiating these pressures, the FIST team, after replanning its efforts, compromised and agreed to January 1997 (nine months hence) as the date for going live with the new North American system alias first FIST pilot.

At the same time Singapore started the process of looking at their requirements. They expected that the FIST team would do this and were quite concerned when the FIST team restricted itself to comparing the ETC 'business model' they had adapted for North America with the South East Asia region and found a 90–95% match. Following some heated discussions between the South East Asia regional management and the FIST team, it was agreed that they would go ahead and, "for the time being," carry out an update of their existing application systems. The South East Asia regional general manager later extended the terms of reference to allow, "for the time being," an upgrade of equipment "to run the updated software efficiently." He was also very critical of what he called the "top-down-approach" taken by FIST. With very little influence or participation by the regions, he feared it would be like the other "past failures of the Computer Centre." For all these reasons, the South East Asia region "bulldozed" their proposal for an independent information systems effort through the next executive meeting.

By the middle of 1996, North America was the only pilot site. Time pressure was beginning to take its toll on the style of the team: FIST management was now actively encouraging the narrowest possible user participation in order to deliver a system by the January 1997 deadline. They were now excluding further input from North America and started to drive the requirements specification predominantly from HQ, using the ETC model as the basis for the "engineering" of the new business processes. In reaction to this, at a Finance conference in November 1996, North America and the other regional managers issued a strongly worded memo demanding broadly based involvement, to avoid wasting effort on a system which, they felt, would ultimately not support their business. The FIST manager complained: " The finance conference has attempted to change the rules with regard to FIST. Prior to this, we were responsible for progressing the approach and we would keep the other regions informed. Now it was suddenly 'agreed' that every man and his dog would be involved. The FIST timetable cannot absorb this extra involvement without bursting." A timetable was attached to this memo to the CEO to show that the project would take three times as long and cost five times as much if participation by other regions was to be allowed. The CEO sided with the FIST team and issued a circular mandating that the FIST project be fully supported by everyone.

As the North America pilot project was still aiming for the January 1997 deadline, two parallel activity streams were developing. First priority was given to producing and issuing a Request for Proposal (RFP) for software and hardware, to be used internationally as the base modules for the global system. Simultaneously, and as an afterthought, the CEO was now calling for a more detailed Cost Benefit Analysis.

THE GLOBAL REQUEST FOR PROPOSAL (RFP)

In June 1996 the FIST team started to document the set of requirements for North America, which, as they were based on the ETC model, could be expanded to a global level. To do this, a mixed team from North America and FIST was assembled at the Co-op's HQ.

The RFP was sent to all the regions to comment. These comments together with the reaction by the FIST team are shown below in Table 2.

The RFP, asking for firm quotes for software, hardware and communications technology, was finally issued in late September 1996. Replies were expected in early January 1997, in order for the selection to be concluded and for the FIST team to put together a capital expenditure proposal in February 1997.

After a rapid evaluation, mainly by the FIST team with some North America input, ORACLE was chosen as the main provider for database middleware and, together with DATALOGIX, for applications software. No decision was made on either the hardware or the communications technology proposals other than to exclude IBM and Andersen Consulting. Privately, however, HP hardware and EDS as the main contractor for the worldwide communications network (and possibly as a support management contractor for all local technology) were favored.

Table 2: Reactions by the FIST Team

Concerns and Comments	FIST Reaction/Action
The requirements for the Consumer and Food Service (together approx. 60% of the Co-op's business) are not covered and examples were given;	The comments and examples of the Europe and South East Asia regions were summarised and inserted into the RFP as an addendum;
The January 1997 deadline is unrealistic;	"January 1997" remains unchanged";
Since the systems and technology chosen will become a global standard, the regions want participation in the evaluation and selection process following the RFP;	"Discussion will continue to ensure that we achieve a reasonable balance of regional; involvement without impacting the timetable";
The concept of common systems for [the Group's] "core information systems" is strongly questioned (wide differences in the business) and not accepted by key regional management.	A list of "core information systems " will be prepared for Executive agreement.

Back in North America, the emphasis now switched from requirements analysis to the design of the new system.

COST BENEFIT ANALYSIS

Six main areas of benefits were identified by the FIST team:

1. *Streamlined Information Flows*—in the current flow if information, specifically the Enquiry-to-Cash flow, "there is a high number of interrupt points, few of which add value to the business transaction", substituting this with an electronic information flow will eliminate duplication, but will require "a common business language."

2. *Reduced Business Cost*—elimination of duplication (of data entry, of reconciliations, etc.) and the introduction of management by exception are expected to result in staff savings.

3. *Reduced Information Systems Costs*—common subsystems are supposed to forestall duplication of development efforts, technology watch and reduce the cost of information systems management (by concentrating it in the centre).

4. *Integration of Information Systems with Business Planning*—the common global systems will be an "integral part of the business rather than an external *ad hoc* activity."

5. *Enhanced Reporting*—common systems are expected to make it easier to introduce standard Key Performance Indicator reports, enable easier implementation of Executive Information Systems and would shorten report cycles.

6. *Implementation of Agreed Key Principles for system configuration*—a set of "Five FIST principles" for system design and configuration was issued to all regional management.

The FIST team then estimated the dollar value of the benefits, restricting themselves, however, to the estimation of efficiency benefits. A summary of the estimates of these direct savings values is shown in Table 3.

The major costs of the project were estimated as shown in Table 4.

A Net-Present-Value evaluation of the FIST project showed that over five years it represents an NPV of $10.18m, showing an internal rate of return of some 70% and a payback of about two years.

THE COMMON APPLICATION SYSTEMS ISSUE

Following the publication of the business and functional requirements analysis for North America, the other regions asked for a clarification of what precisely the FIST team meant by 'common information systems,' specifi-

Table 3: Summary of Maintenance

Savings Description	Redundancies	$ Value p.a.
Recurring Savings		
Elimination of jobs within the business operations	100 (central) 65 (local)	$ 8.25m
Elimination of jobs within Information Services	25	$ 1.25m
Reduction of capital/replacement costs		$ 8.51
Total annual savings		**$ 18.01m**
Once-off savings		
Avoidance of replacement costs, central		$ 1.55m
Avoidance of replacement costs, local		$ 4.3m
		$ 5.85m

Table 4: Estimated Major Costs of the Project

Cost Description	$ Value
FIST Project Costs (including the pilot projects in North America and UAE);	$ 6.3m
Streamlined Order-to-Delivery (pilot and prototype)	$ 6.7m
Global roll-out costs	$ 20.9m
Total FIST Costs	**$ 33.9m**
of which:	
Capital Expenditure	$ 14.4m
External (Consultants, etc.)	$13.3m
Internal costs	$6.2m

cally what was meant by common application systems. In response to this a separate exercise was initiated late in 1996 and then carried out by the FIST team at the Co-op's HQ back in Australasia. As a result of this assignment, in February 1997 a list of the core applications was assembled. In it, the core and non-core applications were determined according to the following definition:

- " Core applications are defined as those which organizations participating in FIST must implement in order to
  (a) manage their business to meet required goals and objectives;
  (b) fully support the Five FIST principles of business function placement; and

(c) meet the information needs of other units of the Co-op group of companies."

- Non-core applications are defined such: "The internal workings of processes by which some organizational outcomes are achieved are of no interest outside that organization. When an organization chooses to use an automated application to meet such needs, that application is considered to be non-core, no matter how essential it may be to the delivery of the outcome."

In the ensuing discussion it was argued that the definition of core applications is too wide and particularly points (a) and (b) would apply to any application within any organization.

Figure 10 depicts the intended split between the proposed "Core," i.e., the global standard applications, compulsory for every regional and local office, and the "Non-Core," i.e., local application systems.

The Core and Non-Core applications, together with an indication whether they were part of the standard FIST application packages,[10] are listed in Table 5.

Developing the North America Prototype and Pilot

Having decided upon the ORACLE/DATALOGIX software as the Group's standard application system, the FIST team began with the implementation of the software in the North America region, and very soon encountered serious problems.

The manufacturing and distribution modules would not conform with the business processes they were selected to support. FIST responded with

Figure 10: FIST Core and Non-Core Applications

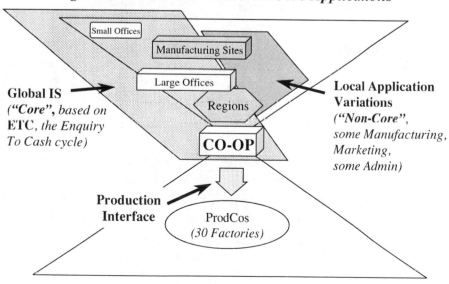

Table 5: Core and Non-Core Applications

Core Applications	Standard?	Non-Core Applications	Standard?
Inventory	Oracle/Dist*	Manufacturing	Datalogix
Purchasing	Oracle/Dist*	Transport	
Import		Marketing	
Planning		Fixed Assets	
Sales	Oracle/Dist*	Treasury	
Export		Project Management	
General Ledger	Oracle/Fin	Office Automation	
Accounts Receivable	Oracle/Fin	Payroll	
Accounts Payable	Oracle/Fin		
Consolidation			
Product Cost Profit & Loss			
Executive Information Retrieval			
* = used to be Datalogix, see below			

setting out a policy that "where a choice existed between the change to business practice or a change to the software system, the business practice would be changed by default." The North America regional manager refused to change business practices which had been developed in response to market and operational reasons. Furthermore, it turned out that software changes needed to be carried out by Datalogix because it transpired that they were going to affect the very kernel of the application. The changes were estimated to cost $1.8m.

However, cost was not the only hindrance to a resolution of the issue. Oracle had just entered into negotiations with Datalogix about acquiring the Datalogix Distribution modules and replacing them with their own ones—so that any investment in Datalogix software would be wasted. For the duration of these negotiations, all adaptation work on the software came to a standstill.

Secondly, because the changes were not just to satisfy North America, but were meant to define the Global Distribution System, it became necessary to address the issue of different numbering systems and coding structures. One area where this was going to be a global difficulty was product and inventory codes. A special subproject was set up to come up with an integrated, international product code schema. This issue is still unresolved. Similar problems plagued the financial software suite. Because they, too, were candidates for a common Global Financial Application Package, a similar exercise was started to create a common system of account codes, suitable for all Co-op subsidiaries. The suggested solution had a 55-digit account number, with built-in logic to reflect the common chart of accounts. The project was abandoned after it was pointed out to the FIST team that in a number of countries (e.g., in most of continental Europe), the chart of accounts is prescribed by fiscal legislation. The use of 'secondary' accounts is a felony in some countries, e.g., Germany.

By September 1997, the FIST project was nine months late and $3.5m over budget. The FIST team suggested to carry out a business-process-reengineering project in North America in order to implement the ETC model there, which would then subsequently become the norm for all the Group's offices:

The BPR project, however, began to go wrong very soon after its inception. The FIST team felt that they had a standard business model (developed at the Co-op's HQ) and that the object of the exercise was to work out ways in which North America's business processes would change so that they fitted that model. North America was expecting a cooperative team project in which they would work out ways in which the North American operation could be improved. This conflict stayed unresolved until well into 1998.

In the meantime, however, North America urgently needed to do some more work on its aging software—three years into the process of trying to get it upgraded. Because their hardware and software were now obsolete, this work took priority and led to a suspension of the FIST project in North America. In early 1998 North America therefore reached an agreement that the software could be altered so that it reflected North America requirements in the first instances.

By the end of 1998, North America had established a stable computer system, using ORACLE financial software to North American specifications. Their regional management got agreement from the CEO that they were no longer considered the pilot project for FIST. The FIST manager summed up the situation in a presentation to the Co-op's executive: "..the North America pilot...originally concentrated on streamlining the business, globally, from North America. One of the things that became obvious was that a global business could not be streamlined from a subsidiary. It needed corporate focus, hence the switch back to HQ. The time in North America was a very useful exercise to prove the software and highlight areas that needed special attention."

The United Arab Emirates (UAE) Pilot

In late 1997, the Co-op decided to open a new office in the Middle East region, in Dubai. This was to support the local trade with the UAE and also help develop more trade into the southern Middle East.

As it became clear that North America was not going to be a satisfactory pilot site, the FIST team decided to select Dubai as the new pilot site to test out the common global system for the Co-op—despite the fact that the UAE office "was only to have about a dozen people and probably does not really have any need to computerise any of its local operations," as stated by the regional manager for Europe, who consequently opted out of the FIST project altogether.

The first installation was going to be the 'standard,' i.e., unmodified ORACLE Financials and Distribution suite and business procedures would be defined around the systems. The first target date for completion was September 1998.

However, for want of adequate local support, the systems could not be developed on site—it was therefore decided to develop the first prototype at HQ, back in Australasia. This delayed the implementation and by April 1999 the prototype was only 50% finished. Continuous difficulties with implementing a computer system in an unsupported and unskilled environment delayed the implementation of the pilot system further—after it had been 'piloted' in Australasia it eventually was handed over as a working system in November 1999 in a reduced form, i.e., the Financials only.

However, it was considered a success and it was intended for use as a model for FIST implementation in other small offices. In 2000, the FIST manager foresaw that Hong Kong, South Africa, the Philippines and mainland China would be next on the list.

Developments Concerning FIST at the New Zealand Dairy Co-Op Head Office in Wellington

The major difficulties with the FIST project began to attract the attention of the CEO. He was especially alarmed about:

- the missed deadlines;
- significant costs (by the end of 1998 between $8m to $ 10m, approximately) without any noticeable benefits; and
- the refusal by North America, followed by Europe, both major regions, to accept the FIST system.

The FIST team persuaded him to become their major sponsor and in March 1999 a revitalized FIST was launched, now with the CEO fully behind it. He had been convinced that the main obstructions to FIST were 'political' and intended not to tolerate any games. The North America regional manager came close to being sacked or demoted because of his refusal to adopt FIST. A regional manager in the European region, who had sent a memo criticising the FIST team for being high-handed, received it back with an invitation to attach his resignation the next time he sends "something like that." Even executives at HQ began to regard criticism of FIST as possible career-limiting moves. In response to this, the regions were distancing themselves from the project, which they now saw as a wholly head-office owned thing.

However, the continuing lack of success and ever-increasing costs of the FIST soon led to a change of mood within the CEO. In November 1999 Ernst & Young was commissioned to evaluate the FIST and the related ETC and other projects. Although noncommittal in the version for public consumption, their report was said to be scathingly critical of both projects, as being overly ambiguous and basically not achievable within the timeframe or project setup in existence.

This proved to be a turning point: the Co-op's executive Board meeting of January 20, 2000 the CEO reassigned the IT portfolio—and with it the FIST project—to the general manager of Finance, who had been an open critic of the project for a long time. Advocating that business reasons should drive the project, he had called for a critical review of the reasons for wanting to spend $35m as early as March 1998.

In rapid succession the FIST project manager as well as the deputy manager for FIST resigned.

Current Challenges

The new FIST manager carried out a post-mortem review of the project. Its major conclusions were that:
1. The importance of strategy alignment between the multinational firm and its information technology function in terms of the Global Business Strategy followed by the enterprise had been seriously underestimated.
2. The requirements for professionalism and in-depth understanding of international issues are essential for even starting an international information systems project.
3. Failing to recognize both had led to unproductive political interaction, which detracted from the objectives of the exercise and ultimately resulted in the demise of the project altogether.

Following these findings, a 'conciliation' committee was formed, consisting of regional business managers and information technology people. After some deliberations, the committee has now commissioned a new, follow-on project—led by the manager of the Europe region—to reposition the thrust of the strategic planning efforts with respect to information technology. The primary objective of the new, rolling Information Technology Strategy will be to improve the effectiveness of the Co-op's operations by contributing to the recognized 'Core Competencies' of the Co-op. The acquisition of any technology to support this will not be an objective in itself but rather a result of the primary objective.

In parallel, the CEO formed an executive project team to review the organizational and authority boundaries between the central, head-office functions and the responsibilities and accountabilities of the regional and local management teams. The outcome of this review is expected to foster a constructive discussion on how best to achieve the balance of central control and local autonomy that the Co-op's global business strategy requires.

FURTHER READINGS

The following are excellent primers into the specific issues surrounding international information systems and the research in this field:

Applegate, L. M., McFarlan, F. W. and McKenney, J. L. (1996). *Corporate Information Systems Management, Text and Cases*, 4th edition. Chicago: Irwin, Chapter 12, 684-691.

Collins, R. W. and Kirsch, L. (1999). *Crossing Boundaries: The Deployment of Global IT Solutions*. Practice-Driven Research in IT Management Series™, Cincinnati, OH, Chapters 4,5.

Deans, C. P. and Karwan, K. R. (Eds.). (1994). *Global Information Systems and Technology: Focus on the Organisation and its Functional Areas*. Hershey, PA: Idea Group Publishing, Chapters 3, 5-8.

Deans, P. C. and Jurison, J. (1996). *Information Technology in a Global Business Environment–Readings and Cases*. New York: Boyd & Fraser, Chapters 5-7.

Laudon, K.C. and Laudon, J. P. (1999). *Essentials of Management Information Systems Organisation and Technology*, 3rd Edition, Upper Saddle River, NJ: Prentice Hall, Chapter 15, 454-491.

Palvia, P. C., Palvia, S. C. and Roche, E. M. (Eds.). (1996). *Global Information Technology and Systems Management-Key issues and Trends*. Ivy League Publishing, Ltd.

Palvia, S., Palvia, P. and and Zigli, R. (Eds.). (1992). *The Global Issues of Information Technology Management*, Hershey, PA: Idea Group Publishing.

Roche, E. M. (1992). *Managing Information Technology in Multinational Corporations*. New York, NY: Macmillan.

A number of cases of less than successful international information systems are contained as the (often historical) international cases in anthologies and monographs on large information systems failure, e.g.:

Flowers, S. (1996). *Software Failure: Management Failure*. New York, NY: John Wiley & Sons, Chapters 3,4-9.

Glass, R. L. (1992). *The Universal Elixir and Other Computing Projects Which Failed*. Bloomington, IN: Computing Trends, Chapters 5,6,9.

Glass, R. L. (1998). *Software Runaways*. Upper Saddler River, NJ: Prentice Hall PTR, Chapter 3.

Yourdon, E. (1997). *Death March: The Complete Software Developer's Guide to Surviving Mission Impossible Projects*. Upper Saddler River, NJ: Prentice Hall PTR, Chapter 5.

ENDNOTES

1 Location is disguised.

2 The case is based on the history of an international information systems project in a real enterprise. It has, however, been simplified, altered and adapted for use as an instructional case. For this reason the enterprise has requested to remain anonymous. Names, places and temporal references have been changed to disguise the enterprise; all monetary references are in U.S. dollars.

3 Includes sub-Saharan Africa.

4 Includes Australia, New Zealand, Polynesia, Micronesia and the Philippines.

5 *Commonwealth of Independent States*, the former Soviet Union.

6 Includes Northern Africa.

7 The following description is substantially simplified in the interest of clarity.

8 Nestle, a large public food company, has annual sales of around $bn 50; Cargill, a privately held food company, has a revenue of $bn 47.

9 Bartlett and Ghoshal (1989) developed a framework for the classification of enterprises operating in more than one country, centred on the level and intensity of global control versus local autonomy. '*Global*' firms maintain high levels of global control while '*Multinationals*' give high local control. '*Transnational*' organisations balance tight global control whilst vigorously fostering local autonomy. This strategy of "think global and act local" is considered optimal for many international operations. '*Internationals*' are an interim state, transiting towards a balance of local and global.

10 i.e., ORACLE and/or DATALOGIX software.

11 The analysis that underlies the categories and relations mentioned in the following sections was derived from the full case, i.e., it draws on an information base that exceeds the details given in the case story in the chapter. It may, however, be used to extend any analysis based on the case as it is contained in the chapter.

12 *Italics* denote the name of a category.

REFERENCES

Butler Cox plc. (1991). Globalisation: The information technology challenge. *Amdahl Executive Institute Research Report,* London, Chapters 3, 5 6.

Ives, B. and Jarvenpaa, S. L. (1991). Applications of global information technology: Key issues for management. *MIS Quarterly*, March, 33-49.

Ives, B. and Jarvenpaa, S. L. (1992). Air Products and Chemicals, Inc: Planning for global information systems. *International Information Systems,* April, 78-99.

Ives, B. and Jarvenpaa, S. L. (1994). MSAS Cargo International: Global freight management. In Jelass, T. and Ciborra, C. (Eds.), *Strategic Information Systems: A European Perspective*. New York, NY: John Wiley and Sons, 230-259.

Jarvenpaa, S. L. and Ives, B. (1994). Organisational fit and flexibility: IT design principles for a globally competing firm. *Research in Strategic Management and Information Technology*, 1, 8-39.

King, W. R. and Sethi, V. (1993). Developing transnational information systems: A case study. *OMEGA International Journal of Management Science,* 21(1), 53-59.

King, W. R. and Sethi, V. (1999). An empirical assessment of the organization of transnational information systems. *Journal of Management Information Systems*, 15(4), 7-28.

Lewin, K. (1952). *Field Theory in Social Science: Selected Theoretical Papers*. Cartwright, D. (Ed.). London, UK: Tavistock, Chapters 5, 8.

Tractinsky, N. and Jarvenpaa, S. L. (1995). Information systems design decisions in a global versus domestic context. *MIS Quarterly,* 19(4), 507-534.

Van den Berg, W. and Mantelaers, P. (1999). Information systems across organisational and national boundaries: An analysis of development problems. *Journal of Global Information Technology Management,* 2(2), 32-65.

APPENDIX: SOME DIRECTIONS FOR THE ANALYSIS OF THE CASE

One way in which a case study can be analysed is to look for the key factors that underlie the dynamics of its story. In sociology such factors are often referred to as the 'categories' of a social scenario and their interplay is termed the 'relations' between them. This terminology is used to suggest some directions for the analysis[11] of this case in the following paragraphs.

The core categories found in the Co-op case fall into two groups, depending on whether the category stems from the business or information technology domain.

In the business domain, six core categories were identified. They are:

1. *Nature of the Business*[12] : the aspects of the Co-op's business that were relevant for the global IS project.
2. *Global Business Strategy*: the relevant aspects of the Co-op's current global business strategy.
3. *Lack of IT at the Headquarters*: the Co-op's inexperience in IT and lack of "IT awareness" culture.
4. *Strategic Migration*: the history of how the Co-op's global business strategy has evolved over the years.
5. *Tradition of Autonomy*: the degree of freedom enjoyed and defended by regional and local management.
6. *Rejection of Global IS*: the actions and manoeuvres to avoid the acceptance of a global standard IS.

The uniqueness of the Co-op's Nature of Business has the most fundamental influence on the case. It determines the essential characteristics of the *Global Business Strategy*. Similarly, the *Strategic Migration* (i.e., the evolution of the Co-op's global business strategy) has a significant impact. All three of these factors combine to establish a distinct *Autonomy Tradition* among the Co-op's local offices and regions. Although assisted by other influences, this *Tradition of Autonomy* is the main determinant of the *Rejection of the Global Information System*, which in itself becomes a major influence in the development of the global system. The other information technology-related category is the *Lack of IT Experience* at the Co-op's headquarters. That naivety with respect to information technology at the Co-op's centre reinforces the *Tradition of* (local) *Autonomy*, this time with respect to the management of local and regional information systems.

There are seven core categories in the information technology domain:

1. *Analysis*: the main assumptions and paradigms governing the analysis of the business requirements for the global IS.

2. *IS Professional Skills*: quality of the professional skills brought to bear on the systems development process.
3. *Domestic Mindset*: inadequacy in the understanding of international issues .
4. *Conceptual Capability*: ability to conceptualise and think through complex issues.
5. *IS Initiative*: the global IS project was initiated by IS, and not from the business.
6. *Global Standard IS*: the standardised, centrist nature of the global IS design ("One system fits all").
7. *IS by Force*: using political power play to force the business to accept the global IS.

A group of four categories shape the character of the remaining three categories in this group. The most fundamental of these 'conditioning' categories is *Conceptual Capability*, i.e., the ability to deal with business complexities. Low conceptual capability has a negative effect on the three other conditioning categories. This is most noticeable for the *Analysis* category, the agglomeration of paradigms used in system design—most of which were fallacious as a consequence. The equally low quality of the *IS Professional Skills* and the pronounced inadequacy to comprehend international issues, summarised in the *Domestic Mindset* category, were further, negative influences on the *Global Standard IS*.

In concert, the four 'conditioning' categories shape the simplistic and inadequate design of the IIS, i.e. the standardised and centrist nature of the international IS encapsulated in the *Global Standard IS* category.

The remaining three categories have direct interactions with the Business Domain:

- Global Standard IS has the strongest interface with the business domain.
- IS Initiative developed as a response to perceived inefficiencies in the international operations of the Co-op; the category also has a strong element of using the IIS as an instrument of central, head-office control.
- IS by Force is the way in which the IT people resort to political power-play as a result of their inability to deal in a rational way with the business people's rejection of the global system.

In addition to the dynamics within each domain, the two domains between them set up a force field in the sense of Lewin (1952). In this force field the conflicting interests of business and information technology functions play out in antagonistic and confrontational interaction. The force field interaction is dominated by the interplay between two key categories. The

Global Standard IS category is the primary causal factor in the *Rejection of the Global IS*. However, both *IS Initiative* and *IS by Force* reinforce the rejection, albeit as a secondary influence. The forces acting in that field are of considerable magnitude and eventually engaged the opposing sides in a cycle of rejection and reaction, which in the end proved strong enough to stop the FIST project altogether.

The *Global Standard IS* as the root cause of these confrontations, however, has another, deeper cause. The inappropriateness of the FIST system's unbending standard design for all regions and local offices of the Co-op, without regard for their significant differences in size, business culture, markets or strategies, is an outcome of the FIST team's erroneous interpretation of the wishes of the CEO. They translated the thrust for increased global co-ordination (to achieve a global business strategy for the Co-op that better balances central control with regional freedom) as a return to full central control, regressing to strategy stance the Co-op had taken in its earlier history. The regional business people correctly interpreted the FIST project as an attempt to roll back their autonomy, and resisted it strongly. Because they suspected the CEO of covertly backing this regressive move, they extended their resistance beyond the *Rejection of the Global Information System* to resistance also against the CEO's thrust towards more regional autonomy—albeit balanced by an equal amount of central control over global concerns such as branding and global product strategies.

The dynamics of the force field also have a further, second-order effect. Three of the main categories, namely *Rejection of the Global Information System, IS Initiative* and *IS by Force,* subsequently formed a cyclical cause effect loop:

- *Rejection of the Global Information System* intensifies the isolation of the IS people (expressed in the *IS Initiative* category).
- This augments their tendency to implement the *IS by Force*, i.e. attempting to achieve by political means what they could not do by rational co-operation with the business people.
- These political power plays, however, only serve to confirm and increase the *Rejection of the Global Information System* by the business people; this starts the cycle of rejection/isolation/politics all over again.

These cyclical interactions ran their course until eventually the CEO involved an outside element—the consultants—to provide an unbiased, third-party opinion. The cycle was then broken by the finding that FIST in its current form would probably never succeed to reach its objectives.

BIOGRAPHICAL SKETCH

Hans Lehmann is a management professional with some 25 years of business experience with information technology. After a career in data process-

ing line management in Austria and South Africa, he worked for some 12 years with Deloitte's as an international management consultant. Mr. Lehmann specialised in the development and implementation of international information systems for multi-national companies in the financial and manufacturing sectors. His work experience with a number of blue-chip clients spans continental Europe, Africa, the United Kingdom, North America and Australasia. In 1991, he changed careers and joined the University of Auckland, New Zealand, where he focuses his research on the strategic management of global information technology, especially in the context of the transformation to international electronic business. Mr. Lehmann has 57 referred publications in journals, books and has spoken at more than 20 international conferences.

LXS Ltd. Meets Tight System Development Deadlines via the St. Lucia Connection

Geoffrey S. Howard
Kent State University, USA

EXECUTIVE SUMMARY

LXS Ltd., a Toronto software house, has identified high market demand for their proposed new product called Estitherm, a Web-based software tool that supports heat loss calculations for architectural engineers designing structures. Estitherm's development requires sophisticated Java programming skills, however, and the project stalls when LXS is unable to hire enough additional programmers to be able to meet the development deadlines dictated by competition. Through lucky coincidence, LXS' Chief Scientist stumbles onto a pool of Java talent while vacationing on the Caribbean island of St. Lucia. Negotiations follow, a contract is signed and the project is quickly brought to successful completion with the aid of Caribbean programmers, working via the Internet. Similar contract arrangements hold the promise for improved economic conditions in Caribbean nations and can reduce software backlogs for companies in developed nations, but better mechanisms are needed to bring together buyers and sellers of IT services.

BACKGROUND

Operating in Toronto since 1986, LXS Ltd. was founded by Lane Bartlett and David Whitsell, two programmers previously employed by CN Railway. At CN, they had been working on a C-language implementation of a freight tracking system that relied on bar code technology. That project bogged down in overruns and was eventually cancelled, but the system's concepts and algorithms had considerable promise, so LXS was founded to produce and market a version of the rail freight system, which was completed successfully in 1988. The package sold well internationally, and LXS grew rapidly.

By 1996 the firm employed about 75 programmers and another 12 people on the support staff, was generating about $26M (Canadian) annually and had successful product offerings in the railway, trucking and warehouse inventory control application areas. Five years later, sales had reached $47M, but the programming staff had only grown to 90 because of the difficulty of finding trained talent in the highly competitive job market. There were an estimated 950,000 unfilled IT jobs in the U.S., and Canada was experiencing similar skilled labor shortages. LXS had added a handful of Web-based applications to its product portfolio, and had organized as shown in Figure 1, below.

Figure 1: LXS Organization

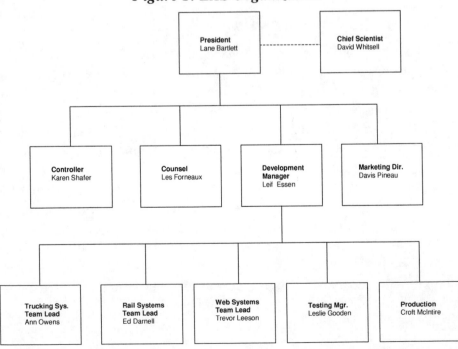

As Figure 1 shows, product development was organized by application areas, with the bulk of the work residing in Ann Owens' Trucking Systems and Ed Darnell's Rail Systems groups. Each group consisted of about 40 programmers, most of whom worked on supporting the successful C-language software packages that accounted for the overwhelming bulk of LXS' revenue. A few of the luckier ones in each group were assigned to designing extensions and refinements for future releases of their packages.

In 1997, the "Web Systems" group was formed to explore Web technology and to develop some small scale product prototypes. LXS had been slow to recognize the potential of HTML technology because Chief Scientist (and LXS cofounder) David Whitsell was skeptical that the Internet would be able to provide the needed bandwidth cost effectively. By early 1997 it was apparent that Whitsell had been too pessimistic, and LXS found itself trying to catch up with the rest of the industry. As part of this catch-up strategy, Trevor Leeson was hired to head the Web group. Leeson had previously been Senior Programming Manager with the Canadian Broadcasting Corporation, which had gone live with one of the first (and best-rated) online Web-based programming guides in the industry. His Web experience included in-depth knowledge of CGI interfaces, PERL and Java, and he was an enthusiastic and visionary cheerleader for Web technology.

Since Leeson was brought into LXS as an outsider, he initially was received coldly, understandably, by the programmers who had been assigned to his new group. Quickly, though, his Scot accent, roaring laugh and sense of humor, and almost nutty enthusiasm for the future of the Web won him respect and cooperation within the group. In addition, he proved himself to be a technical wizard, able to write Java code apparently off the top of his head, with no design support, and make it work right the first time. Nobody else in the group was close to this level of Java ability, so Leeson quickly became a respected leader.

Initially, the new Web group spent all their time attending courses and seminars in order to "tool up" with HTML and Java. They also were sent to an in-depth Windows NT course to understand the architecture, configuration and support of the Microsoft Internet Information Server, as this was the target systems platform for the server-based Internet applications that they proposed to explore for product development.

During this time, Dave Ott, one of Trevor's senior programmers, played a round of golf with Jason Marks, a neighbor and friend. During the round, Marks talked of his problems at work, where he is an architectural engineer for a large engineering design firm. One of the many steps in designing and then obtaining construction permits for commercial buildings requires careful calculation of the thermal properties of the structure. The outside climate, seasonal variation, room dimensions, wall thicknesses and materials characteristics all have a bearing on the heat loss and gain calculations. Each interior space must be carefully studied and complex calculations performed to assure that adequate BTUs and airflow will be available in both heating and cooling seasons. This process is fairly straightforward—the thermodynamics involved are well-understood—but the calculations and analysis are very time consuming. Marks was doubly frustrated because there were at the time, surprisingly, no good PC-based software packages available that automated this design function.

Ott, of course, immediately saw this as an obvious software development opportunity. He arranged to meet Marks the next day for dinner, and they talked further. Marks explained that in the engineering design and construction industries, large design firms such as his compete for contracts to design (but not build) commercial structures. These firms provide a complete array of design services, including design aesthetics, functionality and fitness for purpose, structural loading, survivability, code compliance, electrical and plumbing design, permitting, inspection and HVAC (heating, ventilating and air conditioning) design. The customers of these design firms are contracting companies that actually perform the construction, working under the design guidance of the large engineering firms. The customer contractors range in size from the very small (approximately five employees) to very large firms that are not quite large enough, however, to possess their own in-house engineering design functions.

In the course of the conversation, both Marks and Ott realized that the real marketing advantage of a thermal design software package would derive not from its use in-house by the large design firm. Instead, Marks proposed to almost literally give away the package to the contractors who purchase design services from his firm. This would enable many of these contractors to do initial heat loss estimating on their own, providing their customers with better cost estimates, faster and more accurately. This gesture would serve as a goodwill mechanism that might bring large-scale design business to Marks's company in the same way that giving away pharmaceutical samples to medical offices bootstraps business relationships and contracts. The more they talked, the better the idea sounded, particularly given the void of PC-based thermal estimating software presently available to the engineering design industry.

Dave Ott then prepared a three-page synopsis of this product opportunity and presented it to his boss Leeson, who immediately passed it up the line. Bartlett and Whitsell quickly saw the potential. After only a two-week market-potential study performed for LXS by KPMG, it was clear that the proposed software product was a winner. The proposed package, which Leeson suggested should be called "Estitherm," was quickly approved and funded in November of 1997 as a major product development project for LXS. Further, both Marks and the KPMG consultants suggested that the package should be made available as a Web-based application. Contractors would be able to log onto the site and follow a dialog, entering design specifications for their buildings using sophisticated drag-and-drop graphics, and be able to immediately receive a complete HVAC specification set. In return, the firms making Estitherm available would be building customer goodwill and obtaining contact information for all of the contracting firms that used the Estitherm site.

SETTING THE STAGE

In Toronto, frustrated managers at a software house bite their nails because they have a winning product, plenty of funding, but not enough Java programmers to finish the product and beat out their competitors. Two thousand-five hundred miles south, on the Caribbean island of St Lucia, frustrated managers at a small, new contract programming firm bite *their* nails because they will soon be laying off much of their young Java programming staff for lack of work. What to do? The solution is obvious, but achieving the needed connection between domestic buyers and overseas sellers of software services is anything but easy.

What can be done to eliminate the Information Technology (IT) skills shortage? The inability of companies in the developed nations to find enough programmers to complete their projects is rapidly becoming a strategic emergency (Blumenthal, 1998; PITAC, 1998). This skills shortfall is so severe that it is said (PITAC, 1998) to be constraining the overall growth of the U.S. macroeconomy. Other developed nations such as Canada are experiencing the same shortfalls. Expanded sources of IT expertise must be tapped. LXS Ltd. is no exception. One of their key projects, Estitherm, runs on the Web and enables quick and accurate estimated calculations of the size and type of heating and cooling equipment needed to satisfy a contractor's requirements. Demand for Estitherm is high, but the project is nearly a year behind its original development timeline. What to do?

Meanwhile, economic and development ministers in the small island nations of the Caribbean are struggling to develop stable and growing economies. They must find a way to break away from the rapidly declining plantation-based agriculture of the last century. They must decrease their reliance on unstable tourist income and reverse the brain drain as their best educated and most talented youth flee to Europe and North America for the lucrative jobs that their island homes cannot provide. What to do? Some Caribbean nations have taken tangible first steps to develop offshore IT business. The St. Lucia National Development Commission has built a 20,000-square-foot IT incubator facility to house programmers and provide advanced software development tools and high-speed Internet access, but most of the facility sits idle, wanting for contracts. How can more business be generated?

CASE DESCRIPTION

The Product Team
The Estitherm product development team was formed rapidly. Since it was to be a Web-based product, the team became part of Trevor Leeson's group. Six programmers were assigned into various roles, as shown in Figure 2.

Estitherm Architecture
Leeson initiated the Estitherm project with a series of informal team brainstorming sessions soon after the last programmers had returned from their training courses. Initial discussions centered on overall architecture, and the team decided quickly that most of Estitherm would be written as a Java applet. This meant that most of the Estitherm program code would be

Figure 2: Organization of the Web Systems Team

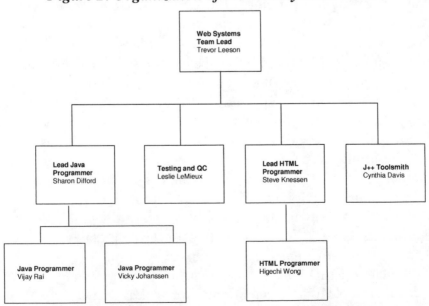

transferred from the server to the client's machine when the application was initially invoked on the Web. After the transfer, the program would run on the client's machine, freeing busy central servers to attend to other tasks. Java was especially appropriate to this task because it is "platform-neutral," meaning that Java applets would run on almost any computer running almost any browser software. During execution of Estitherm on the distant client machine, the program will make data requests via the Internet back to the server to obtain materials properties and climatological data from the Estitherm support database. This rough overall system architecture appears below.

Timetable and Project Management Scheme

As 1998 began and the rugged Canadian winter hit its worst, Team Leader Leeson found himself in a software development manager's "dream" situation. He had a clear project objective, a product that they were confident would be successful in the marketplace, strong executive support from Bartlett and Whitsell and a well-trained team of technically current programmers. Even better, he had the full attention of the team because the rugged January weather eliminated most recreational distractions. They could focus intensively. He drafted an overall project plan, and then modified it in consultation with his staff. The plan called for a working prototype of Estitherm to be ready by September 1998, with all testing and QC complete

Figure 3: Overall Estitherm Architecture

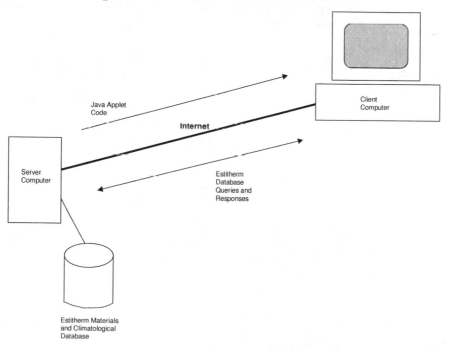

Java Applet Code

Internet

Client Computer

Server Computer

Estitherm Database Queries and Responses

Estitherm Materials and Climatological Database

Figure 4: Estitherm Project Plan

Milestone	Description	Person-Hours	Target Date	Budgeted (CDN) (Labor only)
Detailed Architecture Complete	Design the Estitherm database record layout, the Java applet object structure, and the intercommunication routines.	360	22 January 1998	$16,200
Interface Design Complete	Layout and prototype client screens in the thermal design dialog, then prototype and tune the dialog with user involvement	600	5 March 1998	$28,800
Engineering Algorithms Complete	Engage licensed professional engineer to assist in algorithm design and validation for all thermodynamics module calculations	180	18 February 1998	$15,300
Program Design Complete	Design the internal structure for the Java applet routines	340	20 March 1998	$17,000
Build Server to be Test Bed	Install and test and configure Microsoft IIS on server machine to host testing	160	17 March 1998	$7,200
Applet First Release Coding Complete	Write all first release Java code for the applet and run it in test mode in local browser	1280	17 July 1998	$70,400
Server Database Support Coding Complete	Code all routines necessary to support database lookups on server	410	24 April 1998	$26,650

Figure 4: Estitherm Project Plan (continued)

Support Complete	Design and test backups, recoveries, database maintenance routines, logging and audit routines.	260	8 September 1998	$10,400
Testing Complete	Build 500 test cases for various building designs; validate all cases	600	16 October 1998	$24,000
Beta Test	Release to selected beta testers for comments and corrections (8 weeks in duration)	40	17 December 1998	$1,800
Certification	KPMG Certification Testing and liability validation certification	40	22 December 1998	$16,500
Production Release	Place in production	25	29 December 1998	$1,250
TOTAL				**$235,500**

by late November, and production release at the end of December 1998. The overall plan was submitted to Essen and then to Whitsell, who approved it with minor modifications and authorized the budget as requested. This project management outline appears as Figure 4.

The project plan above reflects labor costs only, and allows for 36% fringe benefits costs above the equivalent hourly rate of the technical personnel involved. KPMG's certification fee was an estimate, and included projected variable expenses above their flat fee for small system certification. In addition to these costs above, another $28,700 was estimated for hardware needed to support the system, and $18,200 for the first-year software licenses.

Team Leader Leeson produced a set of basic project management tracking tools, including a Gantt Chart and CPM diagram. The Critical Path Method graphic was not really necessary because it was obvious to everyone at the start that the first-release coding of the Java applet would be the constraining milestone upon which the entire project depended. Leeson and his boss Leif Essen agreed to hold 30-minute progress briefings each Thursday afternoon for Bartlett and Whitsell. Key technical staff were also to be included in these meetings when their expertise was needed. This simple project management methodology was expected to enable immediate detection of slips in the planned development schedule.

Funding

The funding levels shown in the project plan were presented to Development Manager Leif Essen in mid-December of 1997 and quickly approved by him and President Lane Bartlett. Salary costs would be allocated to the project based on actual hours reported each week. Disbursements for hardware and software would be timed as requested by Team Leader Leeson, with the only requirement being a three-day lead time notification to Controller Karen Shafer.

The Project Begins

Internal Architecture

Work on Estitherm proceeded rapidly and on schedule. Leeson assigned Java programmers Difford and LeMieux to work with HTML guru Steve Knessen to work out the details of the Estitherm architecture. The resulting scheme was as described earlier. All of the climatological data needed to support heat loss and gain calculations would be obtained from U.S. NOAA international databases and formatted and loaded into a Microsoft SQL Server database that would reside on an Estitherm server. This climatological data included worldwide temperatures, humidity, wind and insolation, in all of their seasonal variations. The Server was planned to run Windows NT release 4.0, with Microsoft IIS actually providing Web and database hosting services. SQL Server would use industry standard ODBC (Open Database Connectivity) protocols to support intercommunication with the HTML applications on the clients. The database would also host extensive tables of the engineering properties of materials used in construction. The thermal transmissivity of, for example, an 8 cm thick layer of brick is considerably higher than that of a comparable thickness of pine. Heat loss through concrete floors is much more rapid than if that floor is underlain with a thickness of sand. All database access events would be logged using the audit and journal capabilities built into Windows NT 4.0, thus allowing troubleshooting and bug fixing to proceed rapidly.

On the client side, Estitherm would initially execute by loading HTML code from the server that would present a series of forms to the user. These would be implemented with support from FrontPage, with extensions on the IIS server, and would walk users through a series of input panels requesting initial descriptive data about the proposed building design. After that data had been obtained and validated by editing routines on the server, the Estitherm Java applet would invoke. This highly sophisticated routine would present the user at the client computer with a blank drawing pad and a set of symbols indicating different types of floor, wall and ceiling compositions. The user would then use the mouse to draw out the floor plan of each room in the proposed structure, dragging and dropping materials symbols to each surface after its dimensions had been specified. Microsoft OpenGL standards and tools would be used to build and support these sophisticated graphics routines. As the user specified individual rooms and spaces, those rooms would be shown in a thumbnail graphic at the bottom left of the screen, showing the room-by-room synthesis of the entire structure, level by level. Once the graphic depiction of the structure was complete, the thermal calculations would be executed. The engineering algorithms would be implemented as Java code within the applet on the client's machine, but several queries to the server database would be needed during this process. These queries would provide the needed climatological and materials properties data for the specified structure and its location. This process was anticipated to require about 30 seconds under normal Internet loading conditions, so the design specified provision of a sliding progress bar to keep the user informed.

This architecture design was projected for completion on January 22, but was actually finished on Tuesday the 20th. Leeson reported that happy event in Thursday's update meeting to management, with smiles all around.

On Time, On Budget!

Similar successes were experienced with the Interface Design, Engineering Algorithm Design and Server Build phases of the project. All three of these project subcomponents met their time and cost targets within 5%. Vijay Rai led the interface design. He started with a series of two-hour meetings with contractor and architectural design personnel in Marks's architectural design firm. They had agreed to participate in the design of the product in exchange for a free perpetual license to the finished product. After determining together what the initial design screen should look like, Rai used Visual Basic to quickly build semi-working prototypes of the screens for user reaction, comment and redesign. This process quickly netted a usable, slick interface dialog with which all the users were well-pleased.

Leeson himself took the lead in getting the algorithm design complete because his undergraduate background was in mechanical engineering. He worked for two days with an HVAC (Heating Ventilation and Air Conditioning) consultant from Black and Veatch Ltd. to be sure that he understand the basic heat transfer equations, and finished the project working with Vicky Johanssen over a period of about three weeks. The result was a complete set of validated thermal properties relationships that could then be included into the Java applet. Meanwhile, Cynthia Davis, who was not yet needed on the

Java portion of the project, ordered two high-end PCs, loaded and configured Windows NT 4.0 Server on both machines, and then configured IIS to support the client development and testing. Two parallel systems were created for reliability.

By mid-March of '98, work on Estitherm was proceeding on track, and it appeared that LXS had a winner. Cynthia Davis, the J++ toolsmith, had installed the Java development tool on all four of the Java programmers' PCs, and they had successfully completed several small Java test projects from the technical training course they had recently taken. Coding work on the Estitherm applet began in earnest in late March.

Java R (Not) Us

Java programming started out beautifully. Initially, the team focused on writing and testing code to extract user input from the forms. Next, the database queries to the server were coded and tested, and all went smoothly thanks to the easy interfaceability of the ODBC routines. Trouble came suddenly, however, with the graphics routines. The goal was to initially present the users with a room design panel that started as an empty rectangle. The user could then drag the rectangle's lines in any way necessary to specify the desired shape of a room, and the dimensions would move alongside each line dynamically. After the layout for a room was complete, the user would use simple mouse manipulations to specify the construction characteristics of all surfaces.

Programming these graphics routines in Java proved much more difficult than had been expected. The programmers' learning curve for the difficult vector graphics programming techniques was quite steep. Once the programmers had developed good proficiency, though, the programming process was still very slow because of the inherent complexity of what they were trying to do in the application, and the large quantity of program code necessary to do so. In the April 9 progress meeting, Leeson mentioned that the graphics work was difficult, but expressed confidence that learning effects would enable them to catch up. The following week, in the April 16 meeting, he decided to come clean and confess to Essen and the President that they were two weeks behind at only the third week of graphics programming. No dramatic improvement was anticipated. All concerned had seriously underestimated the difficulty of the graphics programming.

As a possible fix, the entire project team dropped what they were doing and met for two hours on the morning of April 17 to explore whether some different, less graphics-intensive user interface might be employed. This possibility was dispatched quickly, though, because it would require users to enter room dimensions numerically, manually, and made it nearly impossible to account for irregularly shaped rooms. Since Estitherm's goal was to attract customers and win goodwill for the firms who provided it for use on the Web, it was decided that a clunky user interface that angered users would not be acceptable.

Searching for a solution, Essen and Leeson met with all four Java programmers the following Monday. They quickly decided that the only way to finish anywhere close to schedule was to hire more help. The programming task and technology were well understood — they simply needed more hands to get it done fast enough. Since coding productivity was running about one-third of that planned, the three Java programmers needed to become nine programmers. Rather than the projected 1,280 hours needed for this phase of the project, roughly 3,840 hours were needed, an increase of some 2,560 hours. At a labor rate of $40/hour, this would add $101,600 to the projected cost of Estitherm, a 42% cost overrun. These numbers were presented to Bartlett and Whitsell who, to everyone's incredible surprise, agreed immediately. They later confided that they secretly double-budget almost all projects, based on long and rugged experience with software project estimating failures, and were very confident that Estitherm would be a success in the marketplace. To them, a 42% overrun was a small one.

Into the Marketplace

Essen, then, had received authorization to "add capacity" immediately. He contacted several Toronto consulting firms looking for Java programmers who were also familiar with graphics and OpenGL programming. The skills existed, but the lowest consulting billing rate began at $140/hour CDN, clearly an unacceptable number, and even at that rate, it would be at least a three-month wait until six people could be available. Calls to placement firms followed, with the grim news that Java programmers were simply nonexistent in the marketplace. While the staff of three slogged forward on the graphics programming, making steady but slow progress, Essen continued trying to hire more talent. In desperation, he expanded his search to include newspaper ads in major U.S. and Canadian cities, visits to recruiting fairs at McGill University in Montreal (a strong source of computer scientists) and even discussions with colleges about hiring Java programming interns. Obviously, LXS had run squarely against the IT skills shortage—this problem they had been hearing about in the media was real, quite tangible and was directly frustrating their company's strategic product development plan.

Vacation Time

In mid-April, Chief Scientist David Whitsell left with his wife for their annual escape to warm weather. By this time of year in Toronto, the snow has been on the ground for six continuous months and patience is at an end. Whitsell was worried about leaving in the midst of the Estitherm project crisis, but knew that all that could be done was being done. This year, instead of their usual destination of Key Largo, Florida, Whitsell's wife had booked a week in the Caribbean on the British Commonwealth Island nation of St. Lucia. They had booked to stay at Le Sport resort on the north end of the island, so their arrival at the Hewanorra Airport on the south end afforded them an interesting taxi tour from one end of St. Lucia to the other. The early part of the cab ride met their expectations of a slow, sleepy, palm-covered island paradise — hilly roads that later gave way to lush and incredibly green banana plantations in the central part of the island. Driving through the main city of Castries was also stereotypical, and the slow traffic and street vendors were no surprise. Emerging from the north side of town, though, David was amazed to see what appeared to be the early stages of a developing technological industry. There were commercial computer contractors, networking vendors, and they even passed one secure bunker-like building ringed with barbed wire and peppered with satellite antennas.

St. Lucia Development Initiatives

What Whitsell and his wife viewed on their ride to Le Sport was far from their expected stereotype of a stagnant, backward tropical nation. Table 1 lists some of the computer-related organizations actually operating in St. Lucia in 1998.

The "bunker" Whitsell passed is part of a technology incubator project jointly sponsored by the The World Bank and the St. Lucia National Development Commission aimed at attracting information technology services business to the country (SLNDC, 2000). The following specific initiatives have been taken on St. Lucia to attract IT:

- Construction of a 20,000 sq. ft. facility specially designated as an information processing center. The facility is air-conditioned and can be modified to specification. The structure is divided into four sections, with telecommunication lines all the way up to the doorsteps.
- Negotiations with Cable & Wireless, the St. Lucia telecommunications carrier, to reduce telecommunications rates specific to this industry, resulting in an agreement that rates will be consistent with the more competitive rates in the region.

Table 1: Real-World IT-Related Business Operating in St. Lucia, 1998

ISIS World Wm Peter Boulevard Box 1000 Castries, St Lucia 451-6608	Nicholas Institute of Computer Literacy Cadet St Castries 453-7754 Also at Louisville and Vieux Fort 454-7757
Institute of Self Improvement Systems Ltd John Compton Highway Castries 452-1300	Caribbean Computer Literacy Institute Gablewoods Mall Sunny Acres Box 3097 La Clery Castries 451-3030
Micoud Computer Learning Center 32 Lady Micoud St Micoud 454-0556	
University of the West Indies Mome Fortune 452-6290 452-3866	CES San Souci Box 1865 Castries 453-1444 • Fax 452-1558
MainLANLtd (Network administration, consultation /documentation) PO Box 346 Castries, St Lucia Mainlan@candw.lc	Computer Centre Ltd Hill Twenty Babonneau Box 1092, Castries 453-555 FAX 450-6199
Business & Technical Services Ltd GBTS Ltd 49 Mary Ann St Box 1829 Castries 452-4564 FAX 453-1727	University of the West Indies School of Continuing Studies (UWIDITE) Box 306 Castries 452-4080

- Identification of schools interested in training personnel in the applications needed for the IT industry. In addition, St. Lucia has established a government-subsidized training center and maintains a database of potential employees so that they can be easily identified.
- New legislation has been passed to facilitate easy set up of Information Services-related businesses.
- SLNDC has performed identification of local individuals and companies interested in joint venture partnerships with potential investors.

This project was partly funded by $6 million of financing targeted at telecommunications infrastructure improvement in the Eastern Caribbean (Schware & Hume, 1998; The World Bank Group, 1998). The grant project (Schware & Hume, 1998) included funding for vouchers to partially fund training of selected qualified students. It is clear from this project that St. Lucia understands fully the potential of Offshore Programming(O.P.), has elected to invest significantly in creating attractors to industry and has chosen to attempt to capitalize on the seed investment from this grant.

St. Lucia's project is part of a larger multi-nation effort to attract investment to the Eastern Caribbean that has been spearheaded by the Eastern Caribbean Investment Development Service (2000). Headquartered in Washington DC, this agency of the Organization of Eastern Caribbean States promotes offshore business, and information processing specifically, for Anguilla, Antigua and Barbuda, British Virgin Islands, Commonwealth of Dominica, Grenada, Montserrat, St. Kitts and Nevis, St. Lucia, and St. Vincent and the Grenadines. ECIPS emphasizes political stability, quality labor force, English-speaking tradition, proximity to the U.S. and alignment with U.S. time zones, offering a range of incentives including tax holidays and duty-free entry of equipment and raw materials.

Whitsell and his wife enjoyed a couple of days of doing nothing on the beach. But as a high-energy CEO-type, he quickly grew bored and rented a car from the resort so he could try to satisfy his curiosity about the IT activities he saw in town. One thing led to another, and he found himself the next day meeting with a government vice-minister of technology, who explained the incubator project, with a focus on St. Lucia's desire to cultivate an offshore programming industry. The minister explained that the overall economic picture in the Caribbean is one of stagnation or of very slow growth. Most of the countries exhibit "economic dualism," where a modern economy is superimposed upon a less advanced system held over from plantation days. Oil and sugar prices are low, the cost of maintaining those oil and sugar infrastructures is high, tourism is capricious and not fundamental to economic growth, urbanization is imposing ever-increasing social costs and major investments in manufacturing are not occurring. As a result, nearly all of the educated, ambitious youth of the region are leaving to pursue the superior professional employment opportunities in other parts of the world, most especially in England, Canada and the U.S. For example, in relation to the resident population, the overseas population at the end of the 1980s stood at 40% for both Jamaica and Guyana, 23% for Puerto Rico, 21% for Trinidad and Tobago, 15% for Haiti and 10% for Cuba (Girvan, 1997). There is an almost desperate need to find a way to stimulate economic growth in the Caribbean if the brain drain and downward spiral of these economies is to be arrested. Good IT technology training is available in St. Lucia and other Caribbean nations, but the students can't find work and, understandably, depart to developed nations.

Whitsell was then given a tour of the incubator facility. He was surprised to find very high-speed Internet access, hardware equivalent to that at LXS, student programmers with excellent advanced technology skills (including Java (!)), immediately available programmers and a wage structure less than one-third of that in Toronto. He ended his vacation visit to St. Lucia with a promise to return soon to try to construct a Java programming contract arrangement.

Striking a Contract

Immediately upon returning to Toronto, Whitsell called a meeting of the President, the LXS Development Manager Essen and several members of the Estitherm technical project team. They responded enthusiastically to the prospect of getting assistance from St. Lucian programmers, and literally drew a straw to determine who would be lucky enough to accompany Whitsell and Essen on a trip to St. Lucia. The following week, Whitsell, Essen and straw-winner Sharon Difford, Estitherm specifications in hand, met with several managers and programmers of a small software contracting firm in the incubator on St. Lucia. Contract terms were agreed to and papers signed for a "time-and-materials" arrangement at a rate of $16/hour (CDN). Most coordination would be accomplished via the Internet, and the Java code itself could be sent to LXS electronically. Difford stayed behind one more week to coordinate, as seven St. Lucian programmers began coding work on the graphics portion of Estitherm. One of the programmers, Ernest Millston, was a 15 year-old high school student, and Difford was particularly amazed at his skills and energy.

A Clash of Cultures?

LXS programmers were initially concerned about possible cultural differences between their approach to work and that of the St. Lucians. Their stereotype was that people in "that" part of the world are lazy and move slowly, consistent with the universal "No problem" epithet. Most of the stereotype proved, fortunately, to be incorrect. The St. Lucians were agreeable and responsive to inputs from the Toronto-based programmers. After the first week there was, however, a clear indication that the pace of *everything* on the island is much slower than in the North, and the Canadians had trouble communicating and sustaining a sense of project urgency to the St. Lucians. After a bit of trial and error, Essen found that frequent reminders seemed to work. After each prod, the St. Lucians would accelerate, then on about the third day, again begin to lag. Another prod would yield another surge followed by deceleration. While somewhat frustrating, this arrangement worked and "kept the Caribbean programmers going." In the project postmortem meetings, a key item in Leeson's "lessons learned" list was to put one Canadian in place on the island for the project duration in order to keep the work pace high on a daily basis.

Project Completion

The arrangement worked. Toronto programmers and the St. Lucians divided up the work on the Java graphics modules, working out rough spots via e-mail, Internet chat and an occasional phone call. To the regret of all the Canadian programmers, there was no need to make another coordination trip to the Caribbean—the Internet-based communication was adequate. Java programming was complete on August 7, only about three weeks behind the original schedule, and at a very attractive total cost. Rather than the expected $101,600 overrun associated with using Canadian Java programmers (who were nonexistent in the marketplace), the Java portion was completed with only a $40,960 direct labor cost overrun, thanks to the St. Lucia connection.

The balance of the Estitherm project milestones were achieved close to targets, and the system was converted to production only about a month late.

CURRENT CHALLENGES/PROBLEMS

Offshore programming is not a new practice. Indeed, arrangements wherein U.S. firms contract with software developers and technicians in India, Ireland and Pakistan have been in place since about 1985 (Heeks, 1995). This arrangement is mutually beneficial because it provides much-needed employment in the "offshore" nations, improves their balance of foreign credits and aids customer firms in completing stalled or behind-schedule software projects at attractive labor rates (King, 1999). This "offshore programming" ("O.P.") activity is now greatly facilitated by the Internet because the product itself is information, which can move about the world with no delay and at no cost. Software specifications can be sent to contractors, and the resulting software products sent back to purchasers with complete ease. For example, Levi-Strauss, based in San Francisco, contracts for programming services with Cadland Infotech Pvt Ltd., Bangalore, India (Cadland Ltd., 2000). Offshore programming is place-displacement work at its best. Table 2 summarizes offshore programming activity worldwide.

In the Caribbean, however, the offshore programming industry is small, and is struggling for recognition and a way to build business volume. The most visible offshore programming effort in the Caribbean is centered in Montego Bay, Jamaica. Furman University, Greenville, SC, U.S., is involved with the Caribbean Institute of Technology (CIT), training teachers and leading curriculum design with the objective of building a training infrastructure to produce information technologists in Jamaica (Tracy, 1999). This effort, coordinated via HEART, a Jamaican government agency for technology training, recognizes the potential of offshore programming to stimulate the Jamaican economy, and is pursing that opportunity aggressively. Per capita

Table 2: Dollar Value of Offshore Programming Exports by Nations (Heeks, 1995)

Country	Year of Data	Exports (USD, $M)	Growth Rate
Ireland	1990	185	38%
India	1990	120	34%
Singapore	1990	89	43%
Israel	1990	79	39%
Philippines	1990	51	32%
Mexico	1990	38	30%
Hungary	1990	37	53%
Russia	1993	30	N/A
China	1990	18	43%
South Korea	1990	15	40%
Taiwan	1987	11	48%
Egypt	1994	5	N/A
Argentina	1990	4	N/A
Chile	1990	2	98%
Cuba	1993	1	40%

income for the 2.6 million residents of Jamaica is only $6,000, and the 45% unemployment rate (Davidson, 1999) characterizes the desperateness with which the economy needs opportunities such as O.P. INDUSA Offshore, an Indian software company with offices in Atlanta, is seeking programming customers in the U.S., and an initial contract has been arranged with Realm Information Technologies (Atlanta). Realm is currently contracted for thousands of hours of programming work with Indian software firms, and will shift much of that business to Jamaica as soon as the training is complete and the first crop of Jamaican programmers comes online. Edward's Fine Foods, the Southern Company (an electric utility) and Centris Insurance are also carefully studying signing on as customers (Davidson, 1999). There are other early efforts at building offshore programming activity in Barbados, Trinidad and Tobago, Cuba, Antigua and St. Lucia.

The St. Lucian effort, described earlier, has resulted in construction of a sophisticated telecommunications facility to serve as a nucleus for offshore programming on that island. It is in this facility that work on the fictional LXS programming contract began.

The contact between LXS and St. Lucia occurred through serendipity. Had not Whitsell's wife booked a vacation to St. Lucia, this arrangement would not have come about, to LXS', Estitherm's and St. Lucia's mutual loss. Much is to be gained by finding ways to bring together companies like LXS, who desperately need additional IT skills, with nations of the Caribbean, who possess the skills and badly need the revenues.

What is needed to promote these mutually beneficial offshore programming arrangements? Both LXS and St. Lucia were faced with significant challenges. LXS' strategic product development plans were being frustrated by a scarcity of Java programmers. St. Lucia, and the Caribbean generally, is frustrated by the loss of college-trained technicians to other nations because there is not work for them at home. Clearly, the main driver of O.P. is a supply-demand imbalance on the world IT skills market. One nation, usually a more developed one, has a shortfall of available IT skills, and another nation, usually a Third World one, has a skills surplus. Thus, from a macro perspective, O.P. activity can be viewed as a classic economic market-clearing operation, matching demand and supply at some agreed price. If this supply-demand imbalance ever disappears, the entire economic rationale for O.P. is also likely to disappear.

What factors influence the demand for offshore programming services such as those needed by LXS? What prompts a nation such as St. Lucia to try to establish a supply of offshore programming services? How do these buyers and sellers find each other? How will companies all over the globe benefit from an understanding of the kinds of offshore programming services available in the Caribbean and elsewhere? What systematic methods should be put in place to help customers and providers strike arrangements such as that at LXS?

Figure 5 shows a proposed model of factors and effects that may influence the origination of offshore programming arrangements such as the one that benefited LXS. This is a hypothetical model, deduced from real-world experiences, and is offered to spark discussion about all aspects of O.P. activity.

Demand for programming services must exist in excess of the domestic supply. As mentioned earlier, it is this imbalance that drives O.P. activity. Demand requires a backlog of IT work in the services-purchasing nation, a price differential between domestic and offshore programming services, and strong executive support for the concept in corporations in the purchasing nation. Because many executives are uncomfortable entrusting mission-critical projects to the as yet only a partially developed O.P., industry, a supply of projects that are important but not survival-critical can be expected to heighten the demand for O.P. Finally, projects that are well-specified, tightly bounded in scope and cleanly packaged are much more likely candidates for being contracted overseas.

In situations where either there are few well-packaged projects, or where managers simply do not have time to "shop" the world for O.P. contractors, specification packager intermediary firms can catalyze O.P. activity. These firms, which can be domestic or overseas, intervene in the early stages of an

Figure 5: Proposed Supply/Demand Model for Offshore Programming

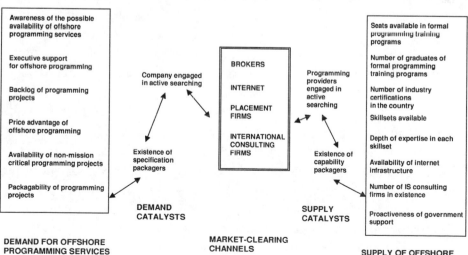

O.P. activity, working with the buyer to clearly package and bound a project and then to find an appropriate overseas contractor. Only a few "specification packaging" organizations exist at present, and as more evolve, O.P. activity can be expected to grow.

At the center of the model are the market-clearing mechanisms that might unite buyers and sellers. At present, much of this is done by large consulting firms, such as KPMG, Accenture and others that are building practice areas in O.P. In addition, a small but growing number of brokerage houses are coming into existence for the sole purpose of pairing buyers of O.P. service together with suppliers. These firms also often assist in the construction of contract terms, the monitoring of project activity and provide troubleshooting support as needed.

On the supply side, one can also envision a small catalyst company that functions as a "capability-packager," fronting for one or more offshore programming firms, or for entire nations. These packagers would have current information on the technical specializations, supply depth and performance records of O.P. firms in supplier nations and thus lubricate the process of locating buyers for services. No such companies have yet been identified, but as O.P. practice grows, the birth of "specification packagers" is hypothesized by the model.

Finally, at the right hand side of the model, supply of O.P. services is obviously a function of the number, quality and productivity of educational programs in IT, the alignment between skillsets needed and skillsets available, connectivity to a particular nation and the attitudes of the host nation governments. Governmental imposition of tariffs on transnational data flows, for example, could greatly inhibit O.P. activity in a particular nation.

Critique this model. What factors in the model seem to be credible? Which do not make sense? What would you add to this model?

The LXS experience was a successful one. What can go wrong in offshore programming activities? List and discuss the pitfalls for both buyers and sellers of O.P. Are these risks strategic, in that they are potential showstoppers for O.P. as a global activity? Are there legal and public policy steps that can be taken to minimize these risks? What agencies, national and global, should take an interest in assuring the success of O.P.?

Research can answer many of the open questions about offshore programming. Work is needed to characterize the kinds of conditions within a software development company that are most likely to prompt a search for an offshore programming solution. More knowledge is needed about the various contracting structures that are employed between O.P. buyers and sellers, and which are most successful. Research on the market-clearing mechanisms are needed—are brokers required, and do brokerage arrangements lead to successful contracts? Are there privacy, tariff, tax or public-policy issues for nations wishing to build O.P. activity? There is no shortage of unanswered questions.

CONCLUSION

LXS' story is typical. Organizations all over North America are experiencing acute shortages of trained information technologists. More than just an inconvenience, these shortages are preventing the timely completion of projects that are often central to strategic organizational goals. In some of these situations, affordable overseas pools of IT talent can be a viable solution, but only if the companies know to look overseas for help. LXS got lucky—they found a source of help in St. Lucia by accident, and were then clever enough to take advantage of that happy accident to complete their project. This is a key message—there is help for beleaguered managers of software projects, but the provider nations must make their availability much more widely known.

IT activity is rapidly "going-global." Today, offshore programming and contract activities such as this LXS project are isolated exceptions. Unless unexpected showstoppers appear, there is every likelihood that systems project teams will routinely and predominantly span national boundaries in the near future. IT managers need to prepare now for the cultural, language, security, economic and quality issues that will accompany the global development teams of the future.

FURTHER READING

CGCED. (1998). Caribbean Group for Cooperation in Economic Development. Workers and labor markets in the Caribbean. *The World Bank Group*. Retrieved on the World Wide Web: http://wbln0018.worldbank.org/external/lac/lac.nsf/c3473659f307761e852567ec0054ee1b/a7df291a495614df852567f20063b23f?OpenDocument.

Davidson, P. (1999). *USA Today*. Jamaica's silicon beach? Retrieved September 13 on the World Wide Web: http://usatoday.com/life/cyber/tech/ctg112.htm.

Girvan, N. (1997). Societies at risk? The Caribbean and global change. Retrieved on the World Wide Web: http://mirror-japan.unesco.org/most/girvan.htm. Paper presented at *Caribbean Regional Consultation on the Management of Social Transformations (MOST) Program of*

UNESCO, Kingston, Jamaica. February.

Heeks, R. (1996). *India's Software Industry: State Policy, Liberalization and Industrial Development*. India: Sage Publications.

Heeks, R. (Ed.). (1995). *Technology and Developing Countries: Practical Applications, Theoretical Issues*. London: Frank Cass & Co. Ltd.

McKee, D. and Tisdell, C. (1990). *Developmental Issues in Small Island Economies*. New York: Praeger Publishers.

Schware, R. and Hume, S. (1998). Organization of Eastern Caribbean States Telecommunications Reform Project. The World Bank Group. Project ID 60PA35730. Retrieved February on the World Wde Web: http://www.worldbank.org/pics/pid/oecs35730.txt.

REFERENCES

Blumenthal, H. (1998). Ready to get your degree in IS? *Netscape Enterprise Developer*. Retrieved January on the World Wide Web: http://www.ne-dev.com/ned-01-1998/ned-01-enterprise.t.html.

Davidson, P. (1999). *USA Today*. Jamaica's silicon beach? Retrieved September 13 on the World Wide Web: http://usatoday.com/life/cyber/tech/ctg112.htm.

Eastern Caribbean Investment Development Service. (2000). Retrieved on the World Wide Web: http://www.ecips.com/descript.htm.

Girvan, N. (1997). Societies at risk? The Caribbean and global change. Retrieved on the World Wide Web: http://mirror-japan.unesco.org/most/girvan.htm. Paper presented at *Caribbean Regional Consultation on the Management of Social Transformations (MOST) Program of UNESCO*, Kingston, Jamaica. February.

Heeks, R. (Ed.) (1995). *Technology and Developing Countries: Practical Applications, Theoretical Issues*. London: Frank Cass & Co. Ltd.

King, J. (1999). Exporting jobs saves IT money. *Computerworld*. Retrieved March 15 on the World Wide Web: http://www.computerworld.com/home/print.nsf/all/990315968E.

PITAC. (1998). President's Information Technology Advisory Committee. Interim report to the *President, National Coordination Office for Computing, Information and Communications*. Retrieved August on the World Wide Web: http://www.ccic.gov/ac/interim.

Schware, R. and Hume, S. (1998). Organization of Eastern Caribbean States Telecommunications Reform Project. The World Bank Group. Project ID 60PA35730. Retrieved February on the World Wide Web: http://www.worldbank.org/pics/pid/oecs35730.txt.

SLNDC. (2000). *St. Lucia National Development Commission*. Information services industry. Retrieved on the World Wide Web: http://www.stluciandc.com/info.htm.

The World Bank Group. (1998). World bank finances telecommunications reform in the Eastern Caribbean. News Release No. 98/1798/LAC. Retrieved June 4 on the World Wide Web: http://www.worldbank.org/html/extdr/extme/1799.htm.

Tracy, A. (1999). Pushing to put Jamaica on the high-tech map. *Business Week Online*. Retrieved July 19 on the World Wide Web: http://www.businessweek.com/bwdaily/dnflash/july1999/nf90719a.htm.

BIOGRAPHICAL SKETCH

Geoffrey S. Howard studies offshore programming in small island nations, telecommuting, computer anxiety and the diffusion of innovation. He accumulated 15 years of industry experience in electrical engineering before coming to Kent, and has published in Decision Sciences, Communications of the ACM, The Computer Journal *and other journals. He was selected twice as one of the top 10 teaching professors at Kent State University, and has been awarded numerous Mortar Board prizes for teaching excellence. He was winner of the Paul Pfeiffer Award for Creative Excellence in Teaching. Dr. Howard is a Registered Professional Engineer in the State of Ohio.*

Success and Failure in Building Electronic Infrastructures in the Air Cargo Industry: A Comparison of The Netherlands and Hong Kong SAR

Ellen Christiaanse
University of Amsterdam, The Netherlands

Jan Damsgaard
Aalborg University, Denmark

EXECUTIVE SUMMARY

Reasons behind the failure and success of large-scale information systems projects continue to puzzle everyone involved in the design and implementation of IT. In particular in the airline industry very successful (passenger reservation) systems have been built which have totally changed the competitive arena of the industry. On the cargo side, however, attempts to implement large-scale community systems have largely failed across the globe. Air cargo parties are becoming increasingly aware of the importance of IT and they understand the value that IOS could provide for the total value chain performance. However, whereas in other sectors IOSs have been very successful, there are only fragmented examples of successful global systems in the air cargo community, and the penetration of IOS in the air cargo industry is by no means pervasive. This case describes the genesis and evolution of two IOSs in the air cargo community and identifies plausible explanations that lead one to be a success and one to be a failure. The two examples are drawn from Europe and from Hong Kong SAR. The case clearly demonstrates that it was the complex, institutional and technical choices made by the initiators of the system in terms of their competitive implications that were the main causes for the systems' fate. The case thus concludes that it was the institutional factors involved in the relationships of the stakeholders that led to the opposite manifestations of the two initiatives, and that such factors should be taken into account when designing and implementing large-scale information systems.

BACKGROUND

Time is the single most important factor in an industry where the distribution of goods moves close to the speed of sound. In the mid-1990s the average shipment time for airfreight was six days. Ninety percent was spent on the ground "waiting" for transport. The need to coordinate and optimize all the ground-based activities in the air cargo community was clear.

Based on weight, air cargo only accounted for 1% of total general cargo transport. However, based on the market value of goods, the share amounted to approximately 25%. Of the total US$200 billion in world-scheduled airline operating revenues, the air cargo industry represented a relatively small share at around US$30 billion.

Just as in other sectors, there was a growing interest in IT in the air cargo community. While most in-house functions had become IT supported and re-engineered in the 1980s and in the early 1990s, the air cargo community was looking beyond organizational boundaries to identify further improvements. Air cargo parties were becoming increasingly aware of the importance of inter-organizational information systems (IOS) and, increasingly, they understood the value that IOS could provide for the total value chain performance.

In the 1990s, many industries had undergone dramatic changes as a result of IT both within organizations and across. However, whereas in other sectors IOS had scored big successes, there were no real signs of deep penetration of IOS in the air cargo community. Although a large number of attempts had been made to automate air cargo processes across stakeholders, there was still no dominant or widely accepted design of an IOS for the air cargo industry that also would satisfy and align the varying demands of the parties involved in air cargo processes.

SETTING THE STAGE

As early as 1975, the International Air Transport Association (IATA[1]) concluded that for 78% of its total travel time, air cargo was at the airport "waiting" for transport and there were no clear signs that there had been much improvement since. According to IATA, this inefficiency was caused mainly by the lack of communication and integration of administrative processes on ground. It was expected that technological innovation such as the development pre-defined open document standards would reduce the waiting time and speed up the time-consuming and error-prone processes such as manual data-entry and re-keying of information. It was also expected that coupling cargo and accounting IS would speed up billing processes, the time to check space availability, bookings and also the reporting procedures.

The following provides more insight into the nature of the air cargo business and the important information flows in this business network.

The black arrows in Figure 1 refer to the physical movement of cargo between the parties in this network while the dotted arrows refer to the information flows between members in this network. It should be clear that the movement of cargo is of a sequential nature and that the information flows can be done in parallel. An example might be the clearing of goods at customs, often cited as a bottleneck by air cargo freight forwarders. The administrative information flows related to customs do not necessarily have to take place in parallel with the physical movement of cargo.

CASE DESCRIPTION

This section provides two tales of electronic infrastructure development. One tale from The Netherlands and one from Hong Kong, both were major hubs in international trade and transportation, and had been important hubs for centuries. See Exhibit 1 for tonnage handled in Hong Kong and at Schiphol airport.

At the beginning of the new century, Hong Kong Special Administrative Region of the People's Republic of China was one of the four largest financial centers of the world, it had one of the three largest seaports in the world, and in 1996 Hong Kong's international airport, Kai Tak, overtook Tokyo's Narita airport in terms of *international* air cargo and became the world's largest. The throughput of Kai Tak was even surpassing its nominated cargo capacity of

Figure 1: Information Flows in the Traditional Intercontinental Air Cargo Chain. (Adapted from Zijp, 1995)

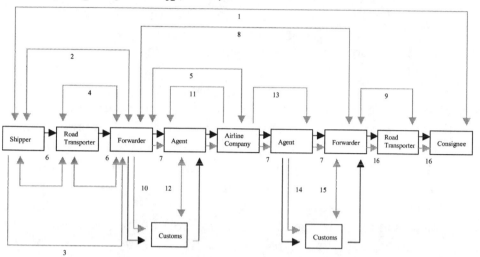

1. Consignee places an order with the shipper and he confirms receipt of the order;
2. Shipper places a transport order with the forwarder and the confirms receipt of the order;
3. Shipper passes on shipping instructions to the forwarder;
4. Forwarder reserves and books freight capacity with the road transporter and he confirms the reservation and booking;
5. Forwarder reserves and books freight capacity with the airline company and he confirms the reservation and booking;
6. Forwarder makes up the bill of lading for road transporter and this document goes with the freight during the road transport;
7. Forwarder makes up an Air Waybill and this document goes with the air freight from one airport to the other;
8. Forwarder gives an assignment to the forwarder at the airport of destination, to reserve and book freight capacity with the road transporter and he confirms receipt of this assignment;
9. Foreign forwarder reserves and books freight capacity at the road transporter and he confirms the reservation and booking;
10. Forwarder supplies information about the air freight sending with the customs and the customs provides the forwarder with the necessary documents;
11. Airline company provides the agent with a booking list for a specific flight;
12. Agent gives information about the load of a specific flight to the customs and the customs gives confirmation to the agent;
13. Airline company provides agent with details about the load of a specific flight at the airport of destination;
14. Agent at the airport of destination gives details about the load of a specific flight to the customs and the customs gives confirmation back;
15. Forwarder at the airport of destination provides the customs there with details about the load and gets information about this from the customs in return;
16. Forwarder at the airport of destination makes up a bill of lading for road transporter and this document goes with the freight during road transport.

1.5 million tons, which emphasized the urgent need for the new airport that was opened at Chek Lap Kok in 1998. The new airport was designed to be capable of handling three million tons of cargo a year, which was expected to be sufficient well into the new century.

While Hong Kong switched from being one of the last remaining British colonies, The Netherlands had established themselves with one of the three largest seaports in the world and served as an important distribution center for cargo into Europe. The relatively close proximity of Rotterdam as a harbor and Amsterdam Schiphol as the airport connected by excellent infrastructure comprised the backbone of The Netherlands' status as Europe's main distribution center. In addition, Schiphol airport was a main hub for passenger travel to and from Europe.

In 1999 the increase in passengers and aircraft movements to Amsterdam Schiphol airport was less than in the preceding years. With growth of 6.6% to 36,772,015 passengers, passenger traffic maintained its market position inside Europe. Freight tonnage remained at virtually the same volume for the second year in succession. The negligible growth of 0.8% was clearly lower than for other European airports. In the year 1999 traffic growth was also affected by the capacity ceiling imposed by the noise abatement measures and the resulting slot allotment system for the airport. Schiphol achieved a freight volume of 1,180,717 tons, significantly less than expected. Its main competitors Frankfurt, London and Paris amounted to 5.7%.

Case 1: Hong Kong: The Traxon Initiative

Four international airlines envisioned that electronic means of coordinating cargo-related information was the key that could provide for more reliable, accurate and timely exchange of information, and eventually a smoother exchange of data that would speed up the processes in the entire air cargo industry.

They took the initiative to form an international electronic network for coordinating transactions between freight forwarders and airlines. They set up three companies; Traxon Asia Ltd., Traxon Europe Ltd. and Traxon World Wide Ltd. Traxon Europe was mainly run by Air France and Lufthansa and Traxon World Wide. Traxon Asia was run by Cathay Pacific, Air Japan and Traxon World Wide. Traxon World Wide played a minor role its main function was to provide coordination between the two regional companies and the four founding airlines.

The content of the Traxon system consisted of three parts: 1) scrolling of flight schedules, 2) making bookings, and 3) status checking for shipments in transit. This seemingly limited functionality, however, provided significant benefits for both airlines and freight forwarders. The airlines benefited from Traxon by getting more detailed information about bookings and an increased number of bookings. They also believed that the use of Traxon reduced the amount of lost business due to busy telephones or absent (busy) sales personnel. The airlines were provided with better information about each booking because the Traxon system reduced the number of errors in the bookings by avoiding manual typing. The freight forwarders improved their efficiency using Traxon because they could find available space with different airlines, make bookings electronically, and therefore they could calculate and quote prices faster and more efficiently. Before Traxon, freight forwarders had to use the phone and call several airlines to find available space. Instead, by using Traxon, they could perform these activities simultaneously. Traxon also allowed freight forwarders to monitor cargo from the air cargo terminal in Hong Kong until it reached its destination overseas.

Traxon was a powerful search, monitoring and booking tool that was carefully designed to meet the requirements of a small niche of the air cargo chain. As such Traxon only supported numbers 5, 7, 10 and 15 of the information flows in Figure 1. It was deliberately designed not to change but to maintain and enforce existing relationships and business structures. The Traxon system consequently did not carry any information about prices or discounts. This left the market opaque for outsiders and preserved the roles and power balance between the airlines and the freight forwarders.

The Implementation Process and Reasons for Its Success

The dilemma for Traxon's system designer was that the airlines would adopt the system insofar as a majority of the forwarders did. At the same time the freight forwarders would only adopt if most of the airlines did. How to get this spiral of self-enforcement going in favor of Traxon was the major challenge.

Thus the systems designers knew that it was essential that all parties would see the benefits of the arrangement, i.e., decide to participate. Traxon was therefore designed to accommodate the needs of the airlines and forwarders, but also to carefully preserve the sensitive distribution of power between them. It was therefore decided not to extend Traxon to any party beyond freight forwarders (for example to shippers or consignees). The Traxon system consequently did not carry any information about prices or discounts. This left the market opaque for outsiders and preserved the roles and power balance between the airlines, the freight forwarders and the shippers.

Another key factor for Traxon's success was that the implementation process took advantage of the respective airlines' local strong holds. So in Hong Kong, the local airline Cathay Pacific was in charge of the local roll out, and similarly in Japan it was the local Japan airlines. A similar approach was applied in Europe and later on in Korea. Furthermore, each local Traxon system had the other shareholder airlines as initial customers, which constituted a significant share of the air cargo market.

After its first years of operation, Traxon was able to enlarge and sustain its position as the dominant electronic trading network provider in Hong Kong's air cargo community. As of January 1998 there were 187 freight forwarding agents connected to the system, resulting in more than 8.8 million messages per year (1997). A number of airlines gave up their defensive actions and they had now joined, which essentially gave Traxon a defacto monopoly in the air freight community in the Hong Kong hub.

Case 2: The Schiphol/Dutch Situation: The Reuters Initiative

In 1992 Reuters, the worldwide press agency and supplier of business information services, started developing an electronic information system on behalf of parties in the spot-markets for air cargo space. The so-called Reuters Initiative was an international initiative with terminals in Amsterdam, London, Paris and Frankfurt. The traditional air cargo community consisted of three broad functional domains: airlines (passengers/cargo carriers as well as cargo-only carriers), ground transport companies (truck and rail transport companies) and freight forwarders that coordinated the door-to-airport and airport-to-door activities at each end.

The content of the Reuters system consisted of three parts: 1) scrolling news page consisting of general and specific air cargo news that can influence the market of air cargo space; 2) summary of all available business information, such as oil prices and exchange rates—this information was helpful in increasing the insight in the influence of these factors; 3) summary of indicative price quotes typed in—these quotes showed to which degree sellers would like to buy or sell space. By changing the indicative prices, contributors could give signals to other parties involved. In this way prices were assigned the function of information carrier. The three parts together provided a complete overview of the spot-markets for air cargo space. Changes could soon be made visible and in this way it was expected that the parties could react to these changes more quickly. Essentially the Reuters' system was designed to support all the information flows in Figure 1.

The Implementation Process and Reasons for Its Failure

The information system ran on trial at the airport Schiphol[2] from August 1993 to January 1994. After that—due to a lack of participation by key parties in the industry—the system was abandoned.

The interactions of customers, forwarders, integrators and carriers were based on the distinction of the two major activities in the business: transport space and shipping services. The market was structured along the lines of the so-called "space-capacity" principle, meaning that carriers provided space and forwarders and integrators provided services to fill capacity. This principle, however, was increasingly contested among the main parties in the market. The airlines were of the opinion that they offered not only space, but also services. The forwarders and integrators, however, maintained the strict distinction of activities in the market between airlines and themselves. Both parties, however, agreed that the market itself was not (yet) a commodity market in which competition is conducted mainly on price. In contrast, the initiators of the Reuters concept claimed that the market for air cargo space could be separated from the market for air cargo service. They claimed that the market for air cargo space had developed itself from a differentiated market to a commodity market. This could be incorporated in the design of the system if:

- it could only contain price information and no product or service information;
- it treated the market as a commodity market where parties could only compete on price instead of in service attributes;
- in the system the sellers (the airlines and the integrators) were explicitly considered sellers of air cargo *space*;

All parties involved were reluctant to participate in the system. At first *the forwarders* had serious doubts, and in the end the support for the system was totally withdrawn. Some of the reasons were: fear for the elimination of the forwarder, fear for decreasing profit margins due to increasing transparency of the market and a general negative attitude towards electronic business. The reaction of *the carriers* was not positive either.

The air cargo market was thus regarded as a commodity market by the Reuters Initiative and thwarted the space-capacity coordination mechanism used by the parties involved. The system did not take these issues into account during the design of the system and that was an important reason for the parties not to participate. As a result of the conflicting interests, the system was abandoned. It was concluded that there is no viability for the system if the parties concerned refused to cooperate.

DISCUSSION

The air cargo market was characterized by a high degree of intransparency, which created substantial market inefficiencies. However, these market intransparencies were in the interests of some of the parties in this market place. Forwarders in particular derive their main reason for their very existence from this lack of transparency. The forwarders acted as brokers and made their living from coordinating the market, and consequently the forwarders had a far more extensive knowledge of the distribution processes than shippers would have. This information asymmetry was clearly in favor of the forwarders and to the disadvantage of shippers. Usually electronic markets favor the buyers and reduce sellers' profits and market power. It is therefore clear that sellers would want to stay away from any system that emphasizes price information. This was not recognized in the Reuters Initiative, whereas in the case of the Traxon system, it was carefully designed to preserve the secrecy of the price-setting process, and therefore Traxon was able to attract the critical number of forwarders to the system whereas Reuters was not.

CURRENT CHALLENGES

What is unique for the diffusion and adoption of these kinds of IOS systems is the combined power of users. If they decide not to join a network, it is devastating for the IOS, as the Reuters Initiative clearly demonstrates. Attracting users is therefore a key requirement for success. For each new individual user, her decision to adopt an IOS creates positive externalities for the other users, because the usability of this IOS increases dramatically with the number of adopters. However this also means, in contrast to many other technologies, that the benefits of being an early adopter can be relatively low compared to be being a "laggard." This is especially true when there are a number of competing and incompatible alternatives present in the market. Thus potential participants of an IOS can effectively block its establishment by simply not adopting the technology.

If the number of adopters reaches a critical mass of users, the diffusion process will self-evolve until saturation is reached and a monopoly is created. Established monopolies are hard to challenge and dissolve, and therefore Traxon had a strong position in Hong Kong. This effect is clear in the two cases. Reuters started with only pre-trade information and no users. The forwarders felt threatened and decided not to adopt which essentially made the system "useless." The Traxon system was owned by four airlines, which also were initial customers of the system, and therefore, Traxon had a substantial share of the market on the supply side to begin with. The forwarders soon followed once they noticed that their position was protected, the system useful, and that a growing number of airlines and fellow forwarders (competitors) were joining the system. Furthermore the Traxon system was first to market, which meant that it did not have to replace any existing and well-established system. Replacing an institutionalized system can be quite a challenge, as the battles between airline passenger CRSs in the U.S. demonstrate.

However, established monopolies may be short lived. At the beginning of the new century, technological innovations on the Internet and their fast adoptions were rearranging the provision of IOS in many industries. They also had a great impact on the international air cargo community. For the IOS owners the change was even more radical because the Internet was eroding their business foundation. The Internet was replacing the service providers as primary means of carrying electronic messages, and innovations in WWW technology were challenging the systems that the IOS providers were offering. The user-friendly interface and the low cost of access to the Internet were also opening the gates to Internet-based IOS for a number of players that earlier could not afford and/or lacked the skills to operate proprietary IOS systems. The advantage of having one IOS system that interconnected all players in an industry segment, as the Traxon system, was being eroded since most players could build and offer their own services on the Internet and most players could access the system over the Internet. An example was non-share holder airlines that earlier were subject to Traxon's de facto monopoly, tempted to launch their own air cargo service on the WWW and reach just as many forwarders. Traxon's future was still uncertain; it had a strong base but was the technological innovation in favor of Traxon's dominant position or did it jeopardize Traxon's attractive position? What necessary steps should Traxon take to maintain its position, and would it be allowed to embark on such a journey by its owners?

APPENDIX

EXHIBIT 1
Million tons cargo handled in Hong Kong and Schiphol airport

	Kai Tak/Chek Lap Kok Airport	Schiphol Airport[3]
2000	1.50	1.22
1999	1.59	1.18
1998	1.58	1.17
1997	1.69	1.16
1996	1.56	1.08
1995	1.46	0.98
1994	1.29	0.84
1993	1.14	0.78
1992	0.96	0.70
1990	0.80	0.60

ACKNOWLEDGMENTS

The authors acknowledge the case contributions of Tonja van Diepen and John Been to the Reuters case. In addition we would also like to thank the many industry participants in Hong Kong and The Netherlands for their time and openness during interviews. This research was in part supported by the PITNIT project (grant number 9900102, the Danish Research Agency) and the Primavera Research Program (http://primavera.fee.uva.nl/) of the University of Amsterdam.

ENDNOTES

1 http://www.iata.com
2 http://www.schiphol.nl
3 Figures from Schiphol statistical annual review 1999. http://www.schiphol.nl/engine/images/stat99.pdf

FURTHER READING

Bakos, J. Y. (1991) Information links and electronic marketplaces: The role of interorganizational information systems in vertical markets. *Journal of Management Information Systems*, 8(2), 31-52.

Besen, S. M. and Farrell, J. (1994). Choosing how to compete: Strategies and tactics in standardization. *Journal of Economic Perspectives*, 8(2), 117-131.

Christiaanse, E. and Huigen, J. (1995). Institutional dimensions in information technology implementation in complex network settings. *European Journal of Information Systems*, 6, 77-85.

Christiaanse, E. and Kumar, K. (2000). ICT enabled coordination of dynamic supply webs. *International Journal of Physical Distribution and Logistics Management*, 30(3-4), 268-285.

Christiaanse, E. and Zimmerman, R. J. (1999). Electronic channels: The KLM cargo cyberpets case. *Journal of Information Technology*, (14), 123-135.

Clemons E. K. and Row M. (1991). Information technology at Rosenbluth travel: Competitive advantage in a rapidly growing global service company. *Journal of Management Information Systems*, 8, 2.

Copeland, D. and Mckenney, J. (1988). Airline reservation systems: Lessons from history. *MIS Quarterly*, September, 353-370.

Damsgaard, J. (1998). Electronic markets in Hong Kong's air cargo community. In Schmid, B. F., Selz, D. and Sing, R. (Eds.), *EM-International Journal of Electronic Markets*, 8(3), 46-49.

Damsgaard, J. and Lyytinen, K. (1998). Contours of electronic data interchange in Finland: Overcoming technological barriers and collaborating to make it happen. *The Journal of Strategic Information Systems*, 7, 275-297.

Damsgaard, J. and Lyytinen, K. (2001). Building electronic trading infrastructure: A private or public responsibility. *Journal of Organizational Computing and Electronic Commerce*, 11(2), 131-151.

Huigen, J. (1993). Information and communication technology in the context of policy networks. *Technology in Society*, 15, 327-338.

Katz, M. L. and Shapiro, C. (1994). Systems competition and network effects. *Journal of Economic Perspectives*, 8(2), 93-115.

Markus, M. L. (1983). Power, politics and MIS implementation. *Communications of the ACM*, 26, 430-444.

McCarthy, D. (1986). Airfreight forwarders. *Transportation Quarterly*, 97-108.

McKenney, J. (1995). *Waves of Change*. Boston: Harvard Business School Press.

Oliva, T. A. (1994). Technological choice under conditions of changing network externality. *The Journal of High Technology Management Research*, 5(2), 279-298.

Robey, D., Smith, L. A. and Vijayasarathy, L. R. (1993). Perceptions of conflict and success in information systems development projects. *Journal of Management Information Systems*, 10(1), 123-139.

Shaw, S. (1985). *Airline Marketing and Management*, London: Pittman.

Short, J. E. and. Venkatraman N. (1992). Beyond business process redesign: Redefining Baxter's business network. *Sloan Management Review*, 7-21.

Wrigley, C. D., Wagenaar, R. W. and Clarke, R. A. (1994). Electronic data interchange in international trade: Frameworks for the strategic analysis of ocean port communities. *Journal of Strategic Information Systems*, 3(3), 211-234.

Zaheer, A. and Venkatraman, N. (1994). Determinants of electronic integration in the insurance industry: An empirical test. *Management Science*, 40(5), 549-567.

BIOGRAPHICAL SKETCHES

Ellen Christiaanse is an Associate Professor of E-Business at the University of Amsterdam. Her major fields of interest include the impact and optimization of electronic delivery channels, supply chains and dotcom start-

ups. She has been awarded several international prizes for her academic work, which was presented at international conferences (ICIS, ECIS, HICSS, Academy of Management) and published in international journals *(*Journal Information Technology, International Journal of Physical Distribution Systems and Logistics Management, European Journal of IS*). She spent almost four years at the MIT Sloan School of Management as a visiting scholar. Dr. Christiaanse has a Master's degree in Organizational Psychology and a Ph.D. in Economics. She can be reached at echristiaanse@fee.uva.nl and her homepage at http://primavera.fee.uva.nl/ .*

Jan Damsgaard *is an Associate Professor at the Department of Computer Science, Aalborg University. His research focuses on the diffusion, design and implementation of networked technologies such as intranet, e-commerce, EDI, Extranet, Internet and ERP technologies. He has presented his work at international conferences (ICIS, ECIS, HICSS, IFIP 8.2. and 8.6)* and in international journals *(*Journal of AIS, Journal of Global Information Management, Journal of Strategic Information Systems, Information Systems Journal, European Journal of IS, Information Technology and People, Journal of Organizational Computing and Electronic Commerce*). Dr. Damsgaard has a Master's degree in Computer Science and Psychology and a Ph.D. in Computer Science. He can be reached at damse@cs.auc.dk and his homepage at http://www.cs.auc.dk/~damse .*

Ford Mondeo: A Model T World Car?[1]

Michael J. Mol
Erasmus University Rotterdam, The Netherlands

EXECUTIVE SUMMARY

This case weighs the advantages and disadvantages of going global. Ford presented its 1993 Mondeo model, sold as Mystique and Contour in North America, as a 'world car.' It tried to build a single model for all markets globally to optimize scale of production. This required strong involvement from suppliers and heavy usage of new information technology. The case discusses the difficulties that needed to be overcome as well as the gains that Ford expected from the project. New technology allowed Ford to overcome most of the difficulties it had faced in earlier attempts to produce a world car. IT was flanked by major organization changes within Ford. Globalization did not spell obvious success though. While Ford may in the end have succeeded in building an almost global car, it did not necessarily build a car that was competitive in various markets. The Mondeo project resulted in an overhaul of the entire organization under the header of Ford 2000. This program put a heavy emphasis on globalization although it perhaps focused too little on international cooperation and too much on centralization. In terms of Ford's own history, the Mondeo experience may not be called a new Model T, but does represent an important step in Ford's transformation as a global firm.

BACKGROUND

An important stream of work in the area of international management (Prahalad & Doz, 1987; Bartlett & Ghoshal, 1989) is concerned with the location paradox: should an internationalizing firm be responsive to local circumstances or go for global integration? On the one hand global integration presents interesting business perspectives, because firms can offer a single product worldwide and use a very uniform way of organizing and producing based on standardized technology. On the other hand there are usually diverse demands being made on multinational corporations (MNCs) by their local customers, host governments or other parties. Managing the location paradox always requires balancing between the local and global perspectives.

In the most basic terms, the advantages of being global are that firms obtain advantages of scale. Imagine if there was really only one global market for a firm, for example if customers demanded precisely the same car everywhere. A firm could build one huge factory from which it could supply the entire world, one marketing center, one R&D unit and so on. The costs per unit of production would be minimal. In reality we, of course, do not have such markets, but there are certain products that benefit from being produced by international firms. Coca-Cola is a global brand and benefits from global advertising. However, the taste of the beverage varies regionally.

Given that most products are not global, surely there are advantages to being local as well. These are best summarized as 'being in touch' with the environment. Firms that operate locally can quicker or better react on customers, deal with local partners and governments and so on. An haute cuisine restaurant usually serves a local custom base and operates locally. Table 1 provides an overview of the advantages of being local and those of being global, as they were conceived by Prahalad and Doz (1986) in their work on the integration-responsiveness grid. All tables can be found in the Appendix.

Because local and global are two countervailing forces, there will always be a tension between the two. Even for fairly global companies, there is a need to act locally (consider what actions Coca-Cola needed to take when people in Belgium got sick due to drinking it) and no local company can completely ignore the forces of internationalization. However, the consequences of this tension for management policy may not be stable over time. Depending on the extent to which firms can unite the global and the local, they are more or less successful in becoming a 'transnational' firm (Bartlett & Ghoshal, 1989). Transnational firms are able to manage the local and the global simultaneously and are thus believed to be able to achieve superior performance. In this light it is interesting to investigate further the consequences of introducing new information technology to a multinational firm. Information technology is thought to be one of the key drivers of globalization. Is IT indeed the stepping stone towards becoming a more global firm? Or, alternatively, does IT simply allow a firm to manage the tensions between the global and the local better, without changing the balance between the two?

This framework will be applied to the case of the Ford Mondeo,[2] a car introduced by Ford in 1993 as a 'world car.' Ford Motor Company barely needs any introduction. It is of course known as one of the world's premier manufacturers of automobiles. Its cars have been sold all over the world for many decades now. Table 2 describes some of Ford's key financial data. A world car is a single car that is sold in different parts of the world, although slight variations may be made to the model. The following three questions will guide the analysis and discussion of this case:

1.   *What were the advantages of going global with its Mondeo for Ford and what barriers did it face to do so?*

Obviously Ford must have thought there were important advantages attached to producing the first-ever world car. These globalization advantages will be discussed in the case in order to get an idea of the strategic motives behind this decision. On the other hand the automobile industry has always faced local constraints, for example in terms of traffic rules, that needed to be overcome. Therefore a delicate balance needs to be found and maintained between going global and operating locally. What kind of managerial challenge did Ford face here?

2.   *Was new IT the key enabler in establishing this global production and supply structure?*

A world car poses new and possibly very different demands upon the organization and technology in use by Ford. Even if the parts going into a world car and the production technology are essentially the same with an ordinary car, a new logistics and communication structure is required to produce the car. From an IT perspective it is especially interesting whether it was the new technology that helped Ford to produce globally or other factors. It has often been suggested that IT is one of the key drivers of the process of globalization. Does the Mondeo case confirm this?

3.   *Has the Mondeo become the new 'Model T'?*

Ford attained much of its fame and present status from the highly successful Model T, a car produced at a very large scale at the beginning of the previous century. The Model T helped Ford to become by far the largest automobile assembler of the world at the time, until its demise in the late 1920s caused a severe disruption to the Ford Motor Company. The world car concept inherent to the Mondeo presented a new mass scale production innovation. Was the performance of the Mondeo good enough to call it Ford's new Model T?

Ford has always been one of the world's largest and most international manufacturers of cars. It was founded in 1903 and first produced abroad in 1904 in Canada and expanded intercontinentally in 1911 to Manchester, England. Chandler (1964) gives a very detailed description of its early history. Ford differentiated itself from its competitors in 1908 through the unique manufacturing strategy implemented by its legendary founder, Henry Ford. Ford decided that economies of scale and a low-cost product would be the key to competitive advantage. Therefore Ford built only one model, the Model T, from 1908 onwards and attempted to do this in mass scale, low-cost production. The reason Henry Ford chose the Model T from his range of designs was that it was most suitable for mass production. The product was fully standardized. One of the innovations Ford introduced was the moving conveyor belt. Demand for the T-Ford grew rapidly, sparked by the low prices and economic growth in the United States. Ford expanded its number of assembly sites across the United States. In 1921 Ford's Model T sold 845,000 units for a U.S. market share of 55%. Ford became a huge industrial corporation over the period, in part because it also integrated backward by acquiring coal mines, railways and steel mills. However, the Model T's success in the end also proved to be its demise. Demand fell steeply after 1921 and in particular during 1926 and 1927 due to the lesser economic situation and increasing substitution by second-hand cars. Ironically the second-hand market was flooded by Ford's own T model. Those consumers that bought new cars were no longer interested in the simple T-Ford model. With these lower volumes Ford was no longer able to maintain its low costs. This initiated a long rebuilding period for the Ford company, which saw its eternal rival General Motors evolve into the world's largest car manufacturer, which it would remain until the present day. GM's Alfred Sloan introduced a number of managerial innovations like the divisional M-form (Chandler, 1964) that provided GM with the ability to produce multiple models and to reconfigure its organization more effectively.

SETTING THE STAGE

In more recent history Ford initiated a new model, which was also seen to be a breakthrough model. Some observers, though not Ford itself, have likened it to the T-Ford. When Ford Motor Company in 1992 publicly launched its plans to produce a world car, it was already its third attempt to do so. The idea behind a world car, sometimes also referred to as a global car, is that one design fits all. More in particular, the efforts by Ford have been aimed at building a car that can at least be mass-sold in both Europe and the United States, by far the largest markets for Ford. The very first attempt by the company to build one single platform that could be sold in different markets all over the globe without major modifications even dates back to 1960 (Kitchen, 1993). This was of course a time when the word globalization had not entered management vocabulary and most car producers were still mainly oriented towards their domestic markets. The project proved not very successful: some 60 days before production was to be started, the U.S. version was cancelled. The reason was that although the car was innovative, being a front-wheel drive economy car, it would also be more expensive to produce than existing larger models. A second try came in 1981 when Ford tried to sell the same Escort model all over the world (Kitchen, 1993). This time a much larger effort was undertaken to design a single model for both markets. Although the Escort in itself proved to be a marketing success, it had little to do with a world car in the end: only two minor parts were identical in the European and North American versions. These two parts were the water pump seal and the Ford oval badge, by the way. This time the main reason was that two distinct development teams had operated simultaneously on both sides of the Atlantic. Each group posed its own idiosyncratic demands. The Ford organization was still not ready at the time, so it seemed.

Under what circumstances did the Ford Mondeo come onto the market? Ford was still a fairly large firm, which was present in all key markets. Especially in Europe and North America, it had established a broad presence and attained a lot of market share. Ford even was European market leader in 1984, but slipped back into fifth place around 1992, just before the introduction of Mondeo. Table 2 gives some market share information for different markets in various years. More recently, after the introduction of the Mondeo, Ford has of course grown through acquisitions. In Europe, the purchase of Volvo in the late 1990s is the most obvious example. However, over the last two decades, Ford also started to invest on a larger scale in Asia. It did so mainly through agreements with Mazda of Japan and Kia of South Korea. In April 1996 Ford even obtained effective control over Mazda. One problem related to both Mazda and Kia, though, was that they were both relatively weak players within their national systems. Kia came close to a bankruptcy in October 1997, after which the Korean government decided to nationalize the company. Mazda has been widely cited as a firm that lacks both scale and bargaining power to be an effective producer on its own. It stands only in fifth place in the ranking of automobile producers in Japan and came close to bankruptcy around 1980. Ford's key financial data are contained in Table 3. They show that Ford Motor Company has grown substantially over the last 25 years, which is in large part due to the external acquisitions and the addition of rental (Hertz) and financial services.

CASE DESCRIPTION

After the 1960 and 1981 failures, Ford started its third attempt to build a world car in 1986. Using the experience of what went wrong in 1981, European and American engineers started designing a new car, under the code name CDW27. Outside suppliers were involved in the project from 1989 onwards to develop specific components and modules of the car in a joint engineering effort. Three different brand names finally emerged, the *Ford Mondeo* for the European market and the *Ford Contour* and *Mercury Mystique* for the North American market. Of these cars, 90% of the elements were identical, although this is hard to see from the outside where the cars appear to be different.

However, certain differences remained. Seat belts and air bags had to be adapted to the local markets. Since U.S. drivers do not always wear seat belts, their cars were provided with larger air bags. European drivers had a smaller, 30-liter air bag. Ford admitted that it had to cope with different supplier processes, which made it tough to achieve the desired component commonality. Furthermore, local conditions and mandates forced a number of changes. Most of the problems arose when Ford had to re-engineer the Mondeo for the North American market and encountered U.S. federal standards and market conditions.

The stakes were high enough for Ford to make the success of this new car crucial. Some $6 billion were invested before it ever came into production, which is far more money than most competitors spend on a new model (the comparable Chrysler Neon cost only $1.3 billion to develop, for example). Because of the radically new concept, it is sometimes referred to as a 'new Model T,' the car that brought Ford its original fame in the 1920s. In Europe, sales of the Mondeo started in 1993; the United States followed some 14 months later. The car was sold in some 76 countries all over the world, although most sales are obviously realized in the United States and Europe.

Motives

Why did Ford decide to try its luck a third time, despite the fact that nobody else in the car industry was building a world car? The answer provided by the company was a reference to its high degree of internationalization, not just in terms of sales, but also in the spread of production sites and R&D knowledge. This led Ford to the conviction that it would be beneficial to consider a global approach instead of a multi-regional or multi-domestic approach. Philip Benton, Ford's President until December 1992, suggested, "A global company can concentrate its resources where they will be used most effectively."

So what advantages did this global structure provide the company with? *Economies of scale* were believed to be the first and most important reason behind the world car project. These economies were not only to be obtained in the production of the different brands, but also in their design and the sourcing of components and parts from third parties. Being able to purchase double the quantities that a normal car model requires obviously gave Ford room for bargaining about prices. A second reason stems from the increased *flexibility* that Ford obtained. Both flexibility in purchasing and flexibility in production are thought to have grown. Ford can switch between locations (Europe and the U.S.) both for its own production as well as for sourcing components from suppliers. It would be easier to cover for delivery deficiencies on either side of the ocean too. Other reasons that were cited less often include *achieving a global image* vis-à-vis customers, *creating new knowledge* through a worldwide network and a *reactive approach* to the loss of market share in some markets. This last point raises an interesting question: did Ford decide to build a world car out of a position of weakness, or one of strength? Although Ford was still clearly the number two manufacturer of cars in the world (after its eternal rival General Motors), Toyota was starting to catch up, as were others. Furthermore, Ford had experienced some pretty bad losses, especially in 1991 when it lost almost $2.3 billion. So the reactive strategy argument seems to have some ground as well, as Ford's position was gradually slipping. Ford felt that it needed to do something new that could again give it a competitive edge over key rivals. Since Ford still had plenty of financial and technical resources available when it embarked on the world car project, it could afford to invest in such a large project. And Ford had the advantages of a strong presence in both the North American and European markets. Ford was strong but getting weaker.

Internal Organization

The Mondeo/Contour/Mystique was built on a project basis, where both the European and North American organizations contributed to the final product. From the earlier adventures with the Escort model, Ford had learned that real integration would be important. When the Escort was designed, two different design teams from Europe and the U.S. were working on it simultaneously. As Benton put it, "When there were opportunities to deviate from the shared engineering plan, both teams made the most of them, protecting their own turf and defending their own ideas about what constituted the 'right' product."

Ford's factories in Europe are concentrated mainly in Germany, the United Kingdom and Belgium. The Ford world car was assembled in three different plants, in Genk (Belgium), Kansas City (Missouri, United States) and Cuatitlan (Mexico). The European plant initially produced some 400,000 units annually and the two North American plants some 300,000 in all. So it may well be concluded that there was an even spread between the two continents.

Some key components in the car were sourced internally. At the beginning of the 1990s, some 50% of components in the automobile industry were sourced internally, but this percentage decreased rapidly. One example of intra-firm sourcing for the world car was the transmission. The manual transmissions were produced in Halewood in the United Kingdom, and Cologne in Germany. The automatic transmissions came from a Ford plant in Batavia, Ohio. This points to a form of regional specialization in the sourcing network, since automatic transmissions are far more popular in the U.S. than in Europe with any new car model. Some 9% of the European Mondeo cars were equipped with automatic transmissions, a figure that was still 3% above Ford's expectations, by the way.

Role of Outside Suppliers

Outside suppliers fulfill a key role in the project, since some $2.5 billion were spent annually by Ford on components and parts for the world car. Important issues arise on the nature of the sourcing network. First of all Ford tried to integrate the European and North American supply bases as much as possible. Albert Caspers, Ford of Europe's chairman before the Ford 2000 program started in 1995, suggests: "The philosophy was to develop a part only once from one supplier in the world. This is the first project where we have done this." One of the key strategies was to reduce the total number of suppliers severely. The Tempo and Topaz models that preceded the American version of the world car had over 700 different suppliers. Ford was able to reduce this number to 227, using a worldwide supply office and early sourcing. The suppliers that participated were chosen through a global search. Ford itself used the term global-capable suppliers to illustrate its require-ments. The suppliers were either chosen on their past performance or on a surrogate part. Dick Fite, who was the CDW27 supply director at the time, says: "The basic management challenge was to bring the two regional supply bases in North America and Europe together to find the best of all worlds in terms of technology, quality, cost and logistic efficiency, so we could rationalize down to the fewest number of suppliers of best-of-class compo-nents on a worldwide scale." One way of achieving this reduction that Ford used was the tiering of suppliers. At Ford in Basildon (UK), Alan Draper, exterior purchasing agent, said (back in 1993): "We have used tiering in areas like instrument panels for several years and are looking to extend the concept to other areas." The suppliers were approached long in advance of actual production. Most of the contracts were agreed upon for a longer period of time. Many suppliers committed themselves to the project around 1989-1990. This allowed Ford enough time to discuss the car and its components extensively with the suppliers.

Just-In-Time is a central element of the production of the world car, although the intercontinental suppliers could, of course, not deliver JIT. For the other supplies, there was a great perseverance in pressing suppliers to set up plants in the proximity of Genk, in the case of the European Mondeo. Ford itself did not hold any stock of components and parts in the plant as part of the JIT system. This is why many new sites were established within 30 km of Genk, delivering within the hour. They included Kautexwerke (gas tanks) and Lin Pac Ekco (interior front door trim panels), who both started production in Belgium, in the towns of Tessenderlo and Overpelt. A second group started production a little further away, such as Ryobi Aluminium Casting. The Japanese parent of this company was asked by Ford to produce transmission and clutch cases for the Mondeo. A new and successful plant was established in Carrickfergus, County Antrim, Northern Ireland. In 1994 it was heralded as the 'best factory in Northern Ireland.' A third track that followed was by suppliers that were already located near Genk. Rehau, from Rehau in Germany, entered into a cooperative agreement with Arrow Molded Plastics of Circleville, Ohio. Together they developed interior scuff plates, which Rehau then produced for the Genk factory and Arrow for North American production. Finally, some European producers moved to North America to establish joint ventures there, as well as Americans coming over to Europe.

The ever-present cost issue played an important role in the sourcing network of Ford. Economies of scale were an important reason to develop a world car. Ford estimated that through the higher volumes, it was able to reduce the cost of supplies by $150 a car. Since some 700,000 cars were made annually, this saved the company up to $100 million a year. The following statement by Draper neatly illustrates the cost pressure that Ford puts on its suppliers: "We are asking our suppliers to absorb all future cost increases resulting from more expensive labor, materials and overhead." Thus these buyer-supplier relationships were not just cooperative, but contained elements of conflict too.

To what extent was this sourcing network international? It involved mainly suppliers that produce in North America and Europe, although some of these suppliers originated from Japan. Of the aforementioned $2.5 billion, $140 million involved exports from Europe to North America and $260 million exports from North America to Europe. The North American share in the components of the European Mondeo was somewhere around 15%. This figure used to be in the range of 1-2% for older models, so this was a really remarkable change. This project also revealed some clear differences between supplier processes in Ford Europe and Ford North America. This created serious problems in the project: achieving maximum component commonality and quality were made much harder. On the other hand it also allowed Ford to gain insight in the peculiarities of the two parts of its organization. These two different practices provided the firm with a possibility for learning.

Information Technology

The Mondeo project posed two different kinds of demands on Ford's information systems. First there was a need for IT to support or replace existing manual labor in the design and engineering area. This is simply a requirement in all modern production, particularly production of automobiles. Because of the increased complexity of cars, the ever-increasing technical demands and cost pressures, all car makers have introduced IT in these processes. Second, Ford was looking at ways to rapidly exchange data between different parts of the world and to support long-distance communication between its employees and with its suppliers. This was specific to the world car project because it put demands on international information exchange that were not there in a regular European or North American project.

The global scale of production allowed Ford to reduce the number of times certain operations had to be performed. Two prime examples of the first kind of IT application mentioned above are structure calculations and design improvements. Ford invested in networked computers for problemsolving in the body structure design. To calculate the optimal body structure, the finite element method is used nowadays. Basically the finite element method calculates what happens when pressure is put on small squares. Up to 70,000 small squares combine to form the body structure of the car. In order to make such calculations, Ford had to use a large and powerful computer. Therefore it bought a new Cray 4MP super-computer during the Mondeo project, which was located at Ford's headquarters in Dearborn, Michigan. This computer served both the European and U.S. versions and ran for almost a year to complete all calculations. Obviously, this kind of application completely relies on computers like the Cray 4MP. The design of the car poses other problems. Fritz Mayhew, chief of North American design of Ford suggested:

"An internationalism has taken over in designs and products, making it much more possible to do a global car." In order to do that, Ford's engineering people had to rely on standardized programs like Computer Aided Design and Computer Aided Manufacturing (CAD/CAM). In 1991 an international engineering team was installed in the Genk plant to prepare for the production launch of the Mondeo. This team exchanged data and pictures with other Ford engineering centers globally. CAD/CAM was the key tool used to reduce development times.

The second kind of IT application mentioned above does not deal with the technical capabilities of computers, but with the ability of IT to support communication processes over longer distances and to integrate geographically remote parts of the Ford organization and its suppliers. During the Mondeo project Ford installed real-time, multi-site, simultaneous engineering and information transfer as well as a global e-mail system. Many up-front investments in facilities were made by Ford to allow for supplier involvement in product development, supply and manufacturing. This included telecommunications and computer equipment. From the earlier adventures with the Escort model, Ford had learned that real integration would be important. To achieve such integration Ford relied more heavily than in the past on information technology, like a complex video conferencing system. Prior to the launch of Mondeo production, video conferencing was already used in communications between Ford's technical centers in Dunton, UK and Metternich, Germany. Later a transatlantic link was established. The video conferencing rooms at Dunton are booked up to 16 hours a day. John Oldfield, head of the world car program, said about the transatlantic video link: "Without video conferencing, the amount of traveling involved and the time differences would make a project like CDW27 near impossible." To make the global engineering project viable, a worldwide communication infrastructure was needed, particularly one that would allow for sufficient communication with external suppliers. However, not everything could be solved by long-distance communication. It was necessary for the project to physically move people. Oldfield, traveled back and forth across the ocean about once a month for six years. Throughout the project there were a minimum of 35 Americans working in the European organization, mostly engineers, purchasing people and finance people. At one point the engineering team consisted of some 800 people. Ford flew hundreds of technicians back and forth across the ocean. Just before production started in Genk, Ford temporarily airlifted some 150 engineers from England and Germany to big, trailer-like mobile offices outside the Genk plant (at an estimated cost of $4 million to $6 million). Their goal was troubleshooting and solving production problems. However, Ford believed it was getting more for its money than the three new models. This includes an improved global communications network. Alex Trotman suggested in 1994 that: "...our investment is in much more than hardware. We've been learning a new way of doing business for the long term. I have envisaged Ford with a global organization since the late 1960s. It's a natural evolution. Now is the right time for such a change. The tools are there—computers and communications—and we have a strong balance sheet. If you make big changes when times are difficult, expediency often takes precedence."

CASE ANALYSIS

1. *What were the advantages of going global with its Mondeo for Ford and what barriers did it face to do so?*

The advantages of going global were demonstrably there. Ford saved money by ordering larger supply quantities. Furthermore it could use the same internally produced parts, such as submissions, for the three cars on both sides of the Atlantic. The case also shows that Ford has struggled to find the balance between global integration and local activities. While the benefits of going global appeared obvious to the firm's managers, Ford was unable to avoid duplicating structures and adapting its cars to local demand. Local regulation was one reason for adapting the cars: North America and Europe obviously differ in some respects. Different consumer tastes also contributed to the adaptations. Europeans and North Americans sometimes tend to use their cars in different ways. For example parking space is limited in most of the (older) European cities and streets can be rather narrow. North Americans often drive longer distances, thus preferring cruise control. Many Europeans prefer manual transmissions because it fits their driving style better than an automatic transmission. Thus some of the barriers to going global could not be overcome by Ford.

2. *Was new IT the key enabler in establishing this global production and supply structure?*

From the case description two arguments stand out. One is that Ford could not have made the transition required for the world car without new means of information technology and communication technology. Second, these new technologies helped to overcome some of Ford's problems, but failed to remove all of its concerns. It was still necessary to move around large numbers of people in order to deal with local production problems for example. Ford seems to have done a good job in integrating some of the technical functions involved in the project, particularly engineering and design. It is also obvious that most, if not all, of the sales efforts were localized.

In fact, most consumers may not have noticed that they were buying a world car. As far as external suppliers are concerned, there is not much information on the use of IT. In historical perspective it seems that what occurred at Ford during the Mondeo project was a change of two kinds when compared with earlier experiences. First, there was information technology to allow for communication across borders, or perhaps we should say across oceans. Second, there was a conscious effort to have employees on both continents communicate with one another about the main design but also about all the details involved in getting the car produced.

3.   *Has the Mondeo become the new 'Model T'?*

Was the performance of the world car project good enough to call this car a new Model T? Ford itself reported to be quite satisfied with the results of the world car project. Sales of the Mondeo model in Europe were quite good from the beginning, 470,000 units over the first 15 months, and it was also chosen as the European car of the year in 1994 right after it was launched. It must be admitted that the first remake of the model came rather quick though, in 1996. Table 5 provides the unit sales of the Mondeo in Europe and its market share.

In the North American market, the sales were reasonable too, although the Model T argeted a smaller segment from the beginning. In North America there were questions surrounding the high pricing, which caused some problems in marketing the product. Ford itself cited the learning effects, both internally and towards suppliers, as a very positive outcome. According to Parry-Jones, the vice-president in charge of the only Europe-based vehicle center in the new Ford 2000 structure: Ford "is now a lot more comfortable with the idea of working across the major regional borders between Ford and its supply bases and between the various organizational elements within Ford." This implies that Ford has increased its ability to conduct such global projects. As such, the company appeared to be quite satisfied with the outcomes of the projects. Although it may not have constructed a new Model T, it did set out in a new strategic direction by becoming a more global firm.

External critics of the project have centered on two issues. The first is whether it is really possible to build a global car and use global suppliers. The problem is that while cost savings drive the need for a global car, there is a danger of the result being too compromised to appeal to any specific market. In other words, consumers in different countries do want special features. Ford encountered this problem for example with the cup holder, that is a standard item in the U.S., but not so in Europe. As has been mentioned before, because of local tastes and regulations, the two versions only have 90% of the elements in common. Some industry watchers have also doubted whether consumers really want a global car. They suggested that an excellent car is what consumers want. Both the Honda Accord and Toyota Camry models have been sold across continents in roughly the same versions as well. But this was not because they had been made with the idea of a global car in mind, but rather because they were built to be excellent cars. These critics suggested that an excellent car can sell globally, but a global car cannot sell without some form of excellence. On a more basic level, one can also wonder whether a car that is produced in only two regions is really global and whether sourcing almost 100% from the same two regions is really global sourcing.

The second issue of criticism concerned the development time of the car. The standard that was set by most Japanese producers is two to three years. It took Ford some seven to eight years to develop the car, and even four years after outside suppliers were first involved. The $150 savings per car that were reported earlier by sourcing in larger quantities were more than offset by a $200 extra investment per car that Ford had to make in the car, following an improved standard that Nissan introduced in the European market in 1991 (including improvements in the suspension and the engine mounts). So the long development time cost Ford dearly.

CURRENT CHALLENGES
Ford Beyond the Mondeo Introduction

After the introduction of the first world car, Ford decided to take the integration of regional organizations further. As part of the Ford 2000 program, it announced in 1994 that the European and North American car businesses would be merged into the division Ford Automotive Operations. The Asian and South American/Rest of World organizations were being left out for the time being. Since January 1, 1995, Ford was organized along product lines, in so-called vehicle program centers. Of these centers, four were based in the United States, whereas one was based in Europe. Each center was responsible for the worldwide design, operations and sales of a single product category. Ford was truly trying to introduce this method of global sourcing in all of its operations. A key statement of the Ford 2000 program was that Ford has "a preference for suppliers with worldwide presence and resources to support global product development and manufacturing strategies." The Ford 2000 program also included centralizing key managerial talent. Finally, it was unclear whether the organization along products in the program vehicle centers according to Ford 2000 would be beneficial. It was reported in the *Financial Times* in 1996 that many motor industry bosses said "Ford has failed to take account of the risks involved in convulsive change and will suffer as a result. Others, however, argue that hesitation today will only make the inevitable task of restructuring more difficult tomorrow." Four years later, in late 2000, reports emerged that the Ford 2000 vehicle program had resulted in a strong centralization of activities in North America. As a result, Ford was thought to have lost touch with its European consumer base, which caused a loss of market share. It was suggested (Muller, Welch, Green, Woellert and St. Pierre, 2000) that the Ford 2000 program led to an overly centralized organization and left Ford without leadership in Europe, South America and Asia. As a remedy the new Ford CEO, Jacques Nasser, reinstalled executives for various regions in 1999. The strong point of the whole Ford 2000 operation and Nasser's subsequent moves appears to be that development times have come down dramatically, towards the level of Ford's main competitors.

The Internet

As far as using information technology is concerned, Ford also took major steps in introducing new tools. The explosive growth of the Internet after the introduction of the Mondeo triggered new opportunities to improve information exchange between Ford and its suppliers. Ford says that its top priorities are currently customer satisfaction and e-business. A much-publicized example is Covisint, a cooperation started by GM, Ford and DaimlerChrysler which aims to be a marketplace for the automobile industry. Much of the data infrastructure of Covisint and other initiatives is taken care of by ANX, the Auto Network Exchange. Ford has participated in ANX since 1998. ANX is a private, virtual network that connects major car makers in North America and more than 280 of their suppliers. It is used among others for design drawings, secure routing of product specifications and EDI transmissions. The advantage of ANX is that it removes existing proprietary connections between buyers and suppliers and thereby improves interchangeability. ANX is much faster than existing communication lines, reducing turnover times by 50 to 75%. This can generate large cost savings, while maintaining or improving the security of data exchange. ANX is able to cope with a large variety of data sources. While exclusive intranets or extranets induce only more connections and a larger burden of work, an open extranet like ANX decreases the number of electronic links. As the number of network members rises, so do the benefits of ANX. Ford's usage of ANX includes CAD/CAM applications, client server applications, interactive mainframe applications and TCP/IP file transfer (for details see: http://www.anx.com/downloads/ford.pdf). ANX and its members have been pursuing expansion outside of North America. As Joe Boyd, telecommunications analyst of Ford in Dearborn, said: "There's the issue of international suppliers needing to get access to applications on servers back here in North America, where we need the flexibility to support ones on other continents. An international ANX would be very desirable to us."

CONCLUSION

To what extent is Ford's experience in trying to achieve global integration by using information technology applicable for other firms and industries? It appears that all firms that internationalize their operations at one time or the other are confronted with conflicting demands. When McDonalds, the icon of global capitalism, internationalized its operations, it soon found out that it was usually necessary to adapt its menu to local demand. Furthermore some countries had regulations that prohibited some of the practices the firm developed in the United States. The benefits of global integration are often taken for granted by internationalizing firms or industry observers. However, there is no such thing as a uniform process of globalization. One may suggest that only 10% of the European Ford Mondeo was different from the North American Ford Contour and Mercury Mystique. However, precisely this 10% raised the cost level of the car to $6 billion and delayed its introduction in North America (Smith, 1994). Even in the Internet age, the location paradox sketched out at the beginning survives.

As for Ford itself it may well be concluded that the Mondeo/Mystique/Contour is a turning point in its history. The world car has fundamentally altered Ford's approach to building cars, which used to be two different approaches, depending on where the car was built. The world car induced an organizational change, in the Ford 2000 program, aimed at globalization. While it is not said that the outcomes of this program are positive, it is an important step in redefining the car industry. Mondeo may not be a new Model T. Then again: will there ever again be a car that bears the significance for mankind that this one model did, with its 15 million units of sales? Perhaps we should forget about the capital T and simply refer to Mondeo as Ford's 'New Model T.'

ENDNOTE

1 The conceptual base of this case study is described in much more detail in Mol and Koppius (forthcoming).

2 The Mondeo was the European version of the car. The North American names are Mystique and Contour. Because the Mondeo was built in the largest quantities, produced and sold earlier, it is generally referred to as 'the world car' by the business press but also by Ford itself. In the remainder of the text,

the name Mondeo will be used to designate the entire world car project (including the North American models).

FURTHER READING

Anonymous. (1994). The world car: Enter the McFord. *The Economist*, July 23, 69.

Bartlett, C. A. and Ghoshal, S. (1989). *Managing Across Borders: The Transnational Corporation*. Boston, MA: Harvard Business Press.

Cavaye, A. L. M. (1998). An exploratory study investigating transnational information systems. *Journal of Strategic Information Systems*, 7, 17-35.

Cleveland, R. (1996). Vive la difference: Ford's Richard Parry-Jones relishes challenges of world cars. *Ward's Auto World*, 32(3), 119.

Mapleston, P. (1993). World car highlights shift in supply relationships. *Modern Plastics*, 70 (10), 52-57.

Mol, M. J. and Koppius, O. R. (forthcoming). Information technology and the internationalization of the firm. *Journal of Global Information Management*.

Prahalad, C. K. and Doz, Y. L. (1987). *The Multinational Mission: Balancing Local Demands and Global Vision*. New York: The Free Press.

Stevens, T. (1995). Managing across boundaries. *Industry Week*, March 6, 24-30.

Threlkel, M. S. and Kavan, B. C. (1999). From traditional EDI to Internet-based EDI: Managerial considerations. *Journal of Information Technology*, December 14, 347-360.

REFERENCES

Anonymous. (1994). The world car: Enter the McFord. *The Economist*, July 23, 69.

Anonymous. (1995). Managing across boundaries. *Business Week*, March 6, 24-30.

Bartlett, C. A. and Ghoshal, S. (1989). *Managing Across Borders: The Transnational Corporation*. Boston, MA: Harvard Business Press.

Chandler, A. D. (1964). *Giant Enterprise: Ford, General Motors and the Automobile Industry*. New York: Harcourt, Brace & World.

Cleveland, R. (1996). Vive la difference: Ford's Richard Parry-Jones relishes challenges of world cars. *Ward's Auto World*, 32(3), 119.

Fleischer, M. (1996). Excellence for the world. *Automotive Production*, July, 12.

Ford Motor Company. (1976). 1975 Annual Report. Dearborn, MI: Ford Motor Company.

Ford Motor Company. (1981). 1980 Annual Report. Dearborn, MI: Ford Motor Company.

Ford Motor Company. (1986). 1985 Annual Report. Dearborn, MI: Ford Motor Company.

Ford Motor Company. (1991). 1990 Annual Report. Dearborn, MI: Ford Motor Company.

Ford Motor Company. (1995). 1994 Annual Report. Dearborn, MI: Ford Motor Company.

Ford Motor Company. (1996). 1995 Annual Report. Dearborn, MI: Ford Motor Company.

Ford Motor Company. (1998). 1997 Annual Report. Dearborn, MI: Ford Motor Company.

Ford Motor Company. (1999). 1998 Annual Report. Dearborn, MI: Ford Motor Company.

Ford Motor Company. (2000). 1999 Annual Report. Dearborn, MI: Ford Motor Company.

Kitchen, S. (1993). Will the third time be the charm? *Forbes*, March 15, 54.

Mapleston, P. (1993). World car highlights shift in supply relationships. *Modern Plastics*, 70 (10), 52-57.

Mol, M.J. and Koppius, O.R. (forthcoming). Information Technology and the Internationalization of the Firm, *Journal of Global Information Management*.

Muller, J., Welch, D., Green, J., Woellert, L. and St. Pierre, N. (2000). Ford: A crisis of confidence. *Business Week*, September 18.

Prahalad, C. K. and Doz, Y. L. (1987). *The Multinational Mission: Balancing Local Demands and Global Vision*. New York: The Free Press.

Smith, D.C. (1994). Kansas City here I come…Ford's 'global car' team gets set for job 1. *Ward's Auto World*, 30(7), 90-92.

Stevens, T. (1995). Managing across boundaries. *Industry Week*, March 6, 24-30.

Vasilash, G.S. (1994). Ford launches its global cars. *Production*, September, 62-64.

LINKS (02-2001)

http://www.ai-online.com/news/120500FordMondeo.htm
http://www.anx.com
http://www.anx.com/downloads/ford.pdf
http://www.covisint.com/
http://www.ford.com/
http://www.just-auto.com/features_print.asp?art=305

APPENDIX

Table 1: The Advantages of Global Integration and Local Responsiveness (Adapted from Prahalad & Doz, 1987: 18-21)

Pressures for Global Integration	Pressures for Local Responsiveness
Multinational customers are important	Customer needs differ
Multinational competitors are present	Distribution channels vary across countries
Investment intensity is high	Substitutes available and product must be adapted
Technology intensity is high	Local competitors important in market structure
There is a high need for cost reduction	Multiple host government demands
Universal market needs	
Access to raw materials and energy is limited	

Table 2: Key Data for Ford Motor Company, 1975-1999. *Source: Ford Motor Company, Annual Reports 1975, 1980, 1985, 1990, 1995 and 1999. Please note that accounting changes may have occurred over this period. Later years include more revenues and income from services. A net loss is signified by - (1980 only). Ford is currently divided in two sectors: automotive and (financial) services. In services key brand names are Hertz and Kwikfit. In automotive Ford owns not only the Ford brand, but also Volvo, Mazda, Lincoln, Land Rover, Jaguar, Aston Martin and Mercury.*

	1975	1980	1985	1990	1995	1999
Sales North America (thousands of units)	3,072	2,457	3,237	3,284	3,993	4,787
Sales rest of world (thousands of units)	1,618	1,969	2,397	2,588	2,613	2,433
Total sales (millions of US $)	24,009	37,086	52,774	97,650	110,496	162,558
Net income (millions of US $)	323	- 1,543	2,515	99	4,139	7,237
Total employees (numbers)	416,120	426,735	369,300	370,400	346,990	364,550
U.S. Employees (numbers)	203,691	189,917	172,200	180,900	185,960	173,064

Table 3: Ford Market Share Between Start of Development of Mondeo and Right After its Launch. Source: Ford Motor Company, Annual Reports 1985, 1995. Please note that for 1985, Europe includes all European markets other than Germany and the United Kingdom. For 1995 its refers to Europe as a whole, including Germany and UK.

	U.S.	Canada	Germany	U.K.	Europe	World
1985	19.0%	17.0%	10.9%	26.6%	8.3	13.7%
1995	25.6%				11.9%	13.3%

Table 4: Short Summary of Events and Their Outcome

Year	Event	Outcome
1960	First attempt to build a world car	American version is never produced
1981	Second attempt, Ford Escort	Two versions differ completely
1986	Third attempt is started	One U.S.-European engineering team
1989	Supplier involvement starts	Many components developed together
1993	Production and sales in Europe	
1994	Production and sales in U.S.	
1995	Ford 2000 program	European and U.S. operations integrated
1999	Ford 2000 program fails	Regional executives re-appointed

Table 5: Number of Units Sold by the Original Mondeo Model and its European Market Share in the Medium-Sized Car Segment

1993	1994	1995	1996	1997	1998	1999
317,765	380,083	353,769	323,727	331,003	317,843	231,943
10.1%	13.3%	13.0%	11.3%	10.8%	9.7%	7.4%

BIOGRAPHICAL SKETCH

Michael J. Mol is from the Rotterdam School of Management, Erasmus University Rotterdam in The Netherlands. He has been a visitor to Temple University in the U.S., WU Wien in Austria and TU Berlin in Germany. His core research focuses on global sourcing strategy. Related research interests are outsourcing, buyer-supplier relations, the influence of the Internet on these relations and the development of business in the European Union. He has taught several courses including European Business and Global Sourcing Strategy. He has published one book, two book chapters and three journal articles, including work in the Journal of Global Information Management *and the* Academy of Management Executive. *He has been a consultant to the UN's International Trade Center (ITC) and various business organizations.*

Lone River Winery Company: A Case of Virtual Organization and Electronic Business Strategies in Small and Medium-Sized Firms

Emmanuel O. Tetteh
Edith Cowan University, Australia

EXECUTIVE SUMMARY

This chapter discusses a case example of Internet infrastructure and e-business strategy management in small and medium-sized firms (SMFs). The case focuses on the key features of electronic business strategy using a virtual organizing framework. Based in the Swan Valley region of Perth, W. Australia, Lone River Winery Co. Ltd., has over the past five years employed the Internet to extend its business scope beyond the Australian wine market. The company produces and markets some of the choicest brands of Australian wines in the UK and Japan using an integrated online ordering and inventory control system. This has saved the company a lot of money and generated sales increases without having to invest in more staff and warehousing facilities. For Lone River Winery (LRW) Co. success on the Net has depended on some critical factors such as in-house technical expertise, continuous search for tools and techniques for managing the firm's core operations, and the continuous reorientation of the online business model in response to changes in the local and international markets.

BACKGROUND

Lone River Winery Co. Ltd. [1] (LRW) is a small family run winery business located in the booming wine industry region of Swan Valley in Western Australia. It was established in 1972 and owned by two members of the management. The CEO of LRW, James Simpson, is an ardent wine expert but also considers himself a serious information technology user since the early 1980s. The company cultivates a total of 12.5 acres in addition to supplies from contract growers. LRW produces a variety of Australian choice wines with main markets in Australia, Japan, UK, Singapore, Malaysia and Taiwan. Its annual production level is about 10,000-15,000 cartons. LRW specializes in Chardonnay, Chenin Blanc, Sauvignon Blanc, Verdelho, Cabernet Sauvignon, Merlot and Shiraz varieties of grapes for its wines. The company employs six permanent staff including three members of the Simpson family. The company also employs about 15-20 part-time staff performing a variety of tasks at the winery (e.g., cellar door wine tasting, wine sales and wine café activities).

SETTING THE STAGE

The importance of the Internet for small and medium-sized firms as a means for developing competitive business infrastructure and as platform for innovative business management strategies is widely acknowledged. This chapter looks at nonagricultural small and medium-sized firms (SMFs) which are classified by the Australian Bureau of Statistics (1999) as having up to 200 employees. These are further grouped by number of employees into micro (less than 5), small (5 to 19) and medium-sized (20 to 200) firms. The Internet infrastructure of a firm is the collection of network technologies, e-business applications, and digital and human resources which enable and/or define the online business operations. Using the Internet, SMFs can build enterprise-wide electronic business (e-business) infrastructures that effectively integrate into those of key market players including suppliers, partners, competitors and customers. One overriding advantage of an Internet-based business infrastructure for small and medium-sized firms is the capacity to extend the scope of their business operations beyond the limitations of their size, resources and competencies (Tetteh, 1999; Tetteh and Burn, 2000). Online small and medium-sized firms adopt a number of generic business formats and strategies in order to benefit from the unique capabilities of the Internet, especially to develop strategic networks, manage market-spanning value chains, add to their core resources and competencies as well as expand their customer bases.

The generic models, broadly categorized as virtual face, virtual alliance and virtual communities, define the orientation of the business in relation to its markets and stakeholder communities including the suppliers, partners, customers (Tetteh and Burn, 2001). In a virtual face orientation, the SMF creates a virtual shopfront, which can enable it to provide general information about the business, products and facilities that allow its key stakeholders (partners, suppliers and customers) to communicate, collaborate as well as purchase and consume its products. The virtual alliance format has all of the previous, but places emphasis on providing a common virtual space for sharing some resources, competencies and customer bases with other firms. In the virtual community model, there exists a large collection of firms employing a mix of the first two formats, with one of them acting as the community owner or manager. One essential feature of virtual communities is that they present a uniform marketspace to the customers and provide divers and a rich set of market-enabling facilities for all. In general, these business models enable an SMF to extend its reach to a larger-than-local market (e.g., regional, national or international) through the global scope of the Internet. The extent of reach will be determined by the strategic agenda of the company as well as practical choices about which markets to serve and how. The business models also enable, to varying extent, the company to extend the range and flexibility of core business operations and interaction with stakeholders.

E-business strategies are needed to consistently align the strategic vision of the firm with the business context and the capabilities of the infrastructure in order to derive maximum benefit from going online. A number of writers have pointed out that the virtual organising approach to business management is especially relevant for doing business online (Venkatraman and Henderson, 1998; Seiber, 1998; Boudreau et al., 1998; Tetteh, 1999). This is because there are intrinsic virtual capabilities associated with the Internet technologies, which need to be understood and internalized by online firms. These capabilities facilitate virtual values such as flexibility of business operations over multiple time-space locations, responsiveness to market changes, agility in the management of interfirm relationships and industry-wide linkages (Rayport and Sviokla, 1995; Goldman et al., 1995; Grenier and Metes, 1995; Venkatraman and Henderson, 1998). Thus the adoption of a virtual organising perspective can provide an effective strategy for an SMF to continuously adapt its environment to the Internet in order to access (to varied extents) resources, skills and markets to suit its strategic needs.

The Internet presents unique features which facilitate virtual engagement and thus shape factors for e-business success including:

1. Substantial capability for market research and information management, and innovative adaptation of industry data into competitive strategy development and management.
2. Extensive range of tools, techniques and solutions for developing enterprise-wide value chains that span markets on local, regional, national and global scales.
3. A dynamic and relatively complex environment involving a growing set of technologies and business solutions which are continuously undergoing upgrade and enhancement in capabilities.
4. Very rich and complex market-making mechanisms which can aid diverse product development and specialization while enabling increasing customer sophistication and participation in business value chain management.

By developing value chain structures that recognize the unique capabilities of the Internet and adopting appropriate business models and strategies, most SMFs can compete in markets where traditionally their small size (measured in terms of number of employees and size of resources base) makes it difficult.

In this chapter, the concept of virtual organization (VO) is used to describe a business entity made up of one of more firms and involving an extensive application of networked information and communications technologies (including the Internet) in support of internal and interfirm business processes. Central to the VO perspective is the active and innovative use of business networks and strategic alliances as a means of generating and managing resources, competencies and increased market access. These collaborative activities among firms are amenable to the capabilities of the Internet. The core feature of successful management of a business on the Internet is the consistent review of the entire business value chain to identify opportunities for adapting and redefining core operations while looking to exploit emerging capabilities of the Internet for creating enhanced business value. This is essentially a virtual organising e-business strategy. Business on the Internet is essentially an information business. The virtual organising strategy focuses on the information and transaction flows that Internet infrastructures enable (Rayport and Sviokla, 1995; Anghern, 1997). The strategy will involve routine review of the relevance of the firm's strategic agenda and the appropriateness of its business model. How these align with the relevant product markets, the demands of the stakeholder community, and existing and available infrastructure capabilities will have to be evaluated continuously in order to achieve a strategic fit (Venkatraman and Henderson, 1998). This is the basis of competitive advantage in the Internet economy.

The E-Business Strategy Model

Small and medium-sized firms need to develop e-business strategies that address their unique features and challenges. SMFs are generally noted as having difficulties in managing business technology. They work with limited access to financial and human resources, which tend to limit their access to global markets and business opportunities. Thus, an effective e-business strategy should address these limitations and facilitate the extension of their business scope over the Internet. In this section, the main elements of a simplified e-business strategy model are described. This model is based on the virtual organising perspective and focuses on five main attributes of online small and medium-sized firms. These five attributes shape the business scope of the online operations of an SMF. Depending on its strategic objectives in going online, the SMF may employ the Internet infrastructure to extend its scope of operations through the transformation of the five attributes. Increased scope in addition to effectiveness and efficiency gains from the Internet infrastructure may, in turn, contribute to performance improvements. The five attributes are:

- Size of resources and technology investments—including number of employees
- Markets and products
- Core business activities and processes
- Inter-firm linkages and market relationships
- Locational diversity of core operations—relating to how flexible the firm manages work over multiple time-space locations

Some of the areas of improvements in performance that may result from successful e-business strategy management include cost savings, increased customer base and sales turnover, improvement in value chain management, better customer relationship management, generation of new product ideas and faster product development, and improved business image.

By paying attention to these attributes, the SMF may select an appropriate business format, based on its strategic objectives in going online. A strategic analysis based on the attributes can help the firm make decisions on which components of the Internet infrastructure to deploy. Using the attributes as a guide, the firm may then follow a five-stage iterative process of e-business strategy development, including:

1. Define appropriate business model.
2. Align strategic goals with virtual business environment and Internet infrastructure.
3. Develop business value chain to exploit opportunities, focusing on informational aspects of the value chain.

4. Manage interactions with market players, including suppliers, partners, competitors, customers and regulators.
5. Continuous reorientation of business (strategic objectives, business model, strategic alliances, markets, infrastructure, etc.) while searching for new values from relationships, and innovative product development.

In subsequent sections aspects of the e-business strategy model will be illustrated with the case of LRW Co. Ltd.

CASE DESCRIPTION

LRW established its online site in June 1995 as a virtual face model, focusing on providing an additional outlet for the firm's operations and to support its strategy to reach international clients. The site has subsequently been redesigned twice to reflect the changes in operation as the organization expanded virtual operations, linkages and strategic alliances and took advantage of a growing e-market for wine.

Product Markets

The company's major markets are in the the UK and Japan where it sells its premium range primarily to restaurants. "These are hard markets," says James, referring to the relatively high price sensitivity and low profit margins in these markets. In Japan, products are sold mainly through mail orders. Difficulties with the UK's complex postal code system has resulted in a lot of frustration for customers and lost sales for the company. Customers who made errors in specifying the delivery postal codes had their parcels dumped in the 'lost goods' section of the post offices. This problem has recently been alleviated by the provision of an integrated online ordering and delivery system. Malaysia, Singapore and Taiwan are smaller export markets.

In Australia, most of the company's sales are in Sydney, NSW, representing about 75% of the total in the country. The company serves its Australian market mainly through wholesale agents and restaurants, and pays commissions on a selective basis. Sales in Perth, Western Australia provided only up to 10%. This reflects James' position that the Western Australian market is 'problematic.' The online operations of the company have not improved sales levels in the home state, given the relative low perception held about the security of online transactions. In order to encourage more sales in Perth, agents receive a higher sales commission of 25%, as compared with 15% in Sydney. "Having a Web presence is very important for our business. We expect things to pick up as the technology becomes more available. Soon people would be able to surf the Internet via their TV sets," says James. The

CEO sees the online business as becoming increasing important, especially as more people gain access to the Internet and use it for shopping.

E-Business Infrastructure

The latest version of the company's Website was commissioned in October 2000 at a total cost of AU$5000. The site is hosted by a local Internet service provider (ISP) via a permanent Internet connection from the company's in-house servers over five 56 kbps modems. Subscription to this service costs the company about AU$400 per quarter. In all, James considers the overall investment in the site redesign and subscription very reasonable. The site features an electronic ordering facility enabling customers to purchase with their credit cards and provides information about wine production and price lists. The company's online shop focuses on marketing products as well as providing corporate and industry information to its clients. The site provides information about LRW's main wholesale agents and sales outlets. The company uses an e-mail list to target advertising. According to the CEO, this provides a simpler and more effective advertising option, yielding about 10% response rate as compared to less than 5% with direct mail and third-party advertising.

Linkages and Relationship Management

LRW maintains extensive linkages with key players in the local, national and international wine industry. It maintains links with a chain of wholesalers and agents of its products in Japan (3 agents), Singapore (1), Malaysia (1) and Taiwan (1). The company has about 14 agents in Australia—Western Australia (1), Victoria (1), Sydney (6+ 5 NSW restaurants), and Darwin Agent (1). The Website also features a page with links to more than 32 wine sites worldwide. There are also links to key local industry players in Western Australia, e.g., Wine Industry Association of WA Inc. and Australian Wine Online.

Impact of Internet Use on Business

The Internet presents opportunities for business scope transformation and can contribute to overall business value. For most SMFs the Internet can facilitate cost reduction, improvements in efficiency and effectiveness in core operations, increased access to wider markets and resources, more flexibility in working mode and multiple time-space management. Table 1 presents a list of some of the advantages, business functions and strategic intents associated with their use of the Internet. The main strategic intents for going online include developing new international markets, forming business

Table 1: Internet Advantage, Applications and Basic Strategies of SMFs (summary list)

Internet Advantage	Business uses/ functions (current and planned)	Basic Strategic Focus
Market exposure	Publishing product information and provide quotes	Improve online business traffic
Wider reach/ global opportunity		Overcome need for more staff
Expand customer base	Communication	Develop new networks/ form alliances
Enable time efficiency/ 24-hr service	Online ordering, booking, registration	Develop web-site to access
Consolidate separate business lines	Online marketing/ direct sales	international markets
Improve business operations	B2B project management	Manage global logistics and
Increase sales/ more sales outlets	Customer relations management	distribution

networks and alliances, improving online business traffic and enhancing global distribution of products.

The CEO considers LRW's e-business a success. He cites in particular the contribution of the online operations to maintaining markets in Japan and the UK. The online infrastructure has also enabled the company to overcome the difficulty of increasing local sales in the WA State by linking up with wholesalers and agents in other Australian states where sales are higher.

The following is a summary of the major impacts of the Internet on the wine business:

- Changes in the way business is done—the business now focuses more on information-intensive processes and looks for ways to enhance them over the Internet.
- Easier access and management of business information—employees and management use the Internet to gather industry-relevant information. The CEO also manages the entire ordering, fulfillment and UK-base warehousing processes from his computer in the Swan Valley.
- Increase in online sales, especially in the UK—over the 1997-2000 period, there has been an increase of about 60% in sales over the Net.
- Emerging opportunity to develop new products—for example in the UK, gains in efficiency in the ordering and distribution processes have opened a window for developing a "gift ordering" product where a customer may place an order for a selection of the wine products to be delivered to a loved one as a gift.
- Increase in customer base—especially in the UK and Japan, the business has recorded improvement in sales through the streamlined ordering and distribution processes.
- Enhance image of the company as a leading player in Australian wines. In the past three years the company has been able to develop a positive

image as one of the best wine producers in West Australia. This has been achieved, in part, through the publicity provided by the Website and targeted marketing with e-mail lists. The CEO mentioned that an increasing number of young people come over to understudy the wine business at LRW. The company's CEO is also actively involved in efforts to develop a new regional brand of wine—the Swan Valley brand. This initiative is in conjunction with a tourism development program for the region.

Factors Contributing Successful Application of Internet at LRW

Successful business operations of an SMF in the Internet economy depend on a number of internal (i.e., firm-specific) factors. These factors include:

- Strategic vision for business scope extension (as held by CEO or manager-owner about the distinctive opportunities provided by the Internet).
- Technical knowledge base (about the Internet both as marketspace and enterprise infrastructure platform).
- Appropriate business model (linking product market with capabilities of the Internet infrastructure).
- Selective application of e-business infrastructure components to support and enhance information processes of value chain.
- Virtual organising e-business strategy (based on a continuous evaluation and adaptation of core competencies, opportunities and markets).

Lone River Winery Co. Ltd. provides a good illustration of how these factors contribute to e-business success.

Technical knowledge about e-business environment. This was found to be an important factor for LRW's e-business success. This supports a common finding in the website reviews of a host of e-businesses where company statements suggest that e-business success depends on the knowledge about the e-business environment held by the CEO or manager-owner. This involved an appreciation of capabilities of e-business facilities, and an understanding of the potential of the Internet for creating business value. In the case of LRW, the CEO of the company mentioned that he has been involved in computers 'since the days of CPM/DOS operating systems.' The CEO with the help of the daughter—an employee with a bachelor's degree in Commerce—takes care of the routine management of the Website. According to him, his technical knowledge helped him in negotiating a good deal with his chosen ISP, and that he had to scan the market and compare prices before settling on the present ISP. Although the website redesign was outsourced, the company considers the ability to make changes on a routine basis in-house very important, and the CEO expressed his eagerness to receive input from clients and visitors on aspects of site usability.

Virtual organising over the Internet requires a continuous reorganization of business processes, infrastructure components as well as product mix in order to respond to changes in markets. Translating strategic decisions into product features, marketing information and online presentation, and introduction of new online facilities require changes to site content and format on a routine basis. This also involves appreciation of the soft issues of the Internet environment, such as customer perception of value online, brands management, and how to capture and sustain the interest of visitors and ensure revisits.

Strategic visioning for the Internet. The ability to articulate a strategic vision about the company's use of the Internet is another important factor of success. This would depend on the knowledge base about e-business environment and is the basis of selecting an appropriate e-business model. In the case of LRW, there is a clear vision to adopt the Internet infrastructure to support the company's export strategy, especially since the local market in WA is very small. The company sees the Internet as a means of advertising its products to other Australian states and to its international targeted markets, especially the UK and Japan.

Appropriate choice of e-business format. Employing a virtual face business model enables the company to reach the potential clients in other Australian states, especially in Sydney, NSW. In the UK, the company's online ordering facilities enable it to effectively manage orders from the customers of its premium product range. At LRW, there are indications of moving the e-business along a virtual alliance model where other wineries in the Swan Valley region would participate in some form of common brand marketing. Such an alliance may be implemented as a common webspace featuring common facilities for the participating group of firms. Another implementation approach would be to just provide visibility for the alliance and its objectives, with smart links on the company's website to the sites of the participating firms.

Selective application of e-business infrastructure. LRW has an ongoing commitment to improve its online systems as part of the company's marketing strategy. Since 1995 the company has undertaken two major upgrades of its e-business infrastructure. LRW has recently joined the London Wine Bonded Warehouse—a facility allowing international wine producers to stock their produce in UK without incurring import tax and sales tax until such time as actual sales of wines are made. Participation in the bonded warehouse further facilitates its ability to check and process orders electronically and relay purchase information directly to the warehouse for fulfillment.

Virtual organizing e-business strategy. Given the fact that LRW is already a major supplier of wine production services to many of the other Swan Valley wineries, it is easy to see the company's e-business developing along a path focusing on new product lines, especially in the business-to-business category targeting the other wineries.

A review of the company's initial website, developed in 1995, shows that LRW has an active strategy for enhancing its online infrastructure. A comparison of the three versions of sites shows radical improvements to site design and functionality. One key difference is that earlier versions of the site attached significant importance to clients visiting the physical location of winery. At that time physical awareness (as associated with traditional marketing strategies) was considered key to image building and brand loyalty by the company. This emphasis has become reduced in the current version of the site. The online ordering facility has remained simple in layout, but the new hyperlinks emphasize stronger alliances along the value chain and a broader commitment to the global marketplace. This has resulted in a considerable increase in national and international sales. The LRW site encourages visitors to join the company's electronic mailing list on wine and health to learn about developments in the industry and the company's products as well as communicate directly with the company. This is aimed at building brand loyalty.

It is clear that the company is exploiting its competencies both in quality wine production and in-depth appreciation of how to utilize the opportunities presented by the Internet for its export market. The company has recognized that the best outcome from Internet use is primarily in cost management of its high premium but low margin export markets. Integrating the company's online ordering facilities with the Bonded Warehouse has enabled LRW to realize huge cost savings and effective management in inventory, export, staffing and product delivery. Now the CEO can manage all sales in the the UK over the Net from the home base in WA. Without the online facility, the company would need to have at least one sales representative in the UK to take care of orders and arrange delivery from stocks in the warehouse. Given the competitive pricing and very small profit margins of this market, such cost savings are crucial for the company. The integrated online ordering-bonded warehouse system also enables the company to monitor stock levels in order to plan future productions. Easy access to real-time stock levels means the company can engage in selective strategy to push sales through its gift services—a new product line aimed at customers who would like to place orders for choice wines for delivery to friends or loved ones.

Continuous reorganization of the critical value chain processes (especially in inventory control, orders management and product delivery), while looking for online solutions to facilitate these, has provided major enhancements to the company's overall performance.

CURRENT CHALLENGES FACING THE ORGANIZATION

LRW Co. Ltd. continues to build on its strengths in e-business management to improve efficiency and effectiveness in its export markets. According to the CEO, the company considers its online presence important as it expects the Internet to become more widely used for shopping in Western Australia. In the near future, attention of the business is to increase the critical mass of its local online customers through participation in a common Swan Valley wine brands initiative. This program will involve an active advertising campaign in the WA region and elsewhere for a Swan Valley brand as part of building a quality standard in Australian wines. It also forms a part of tourist drive for the region.

ENDNOTE

1 The names of the company and persons in the SMF case have been disguised to protect the privacy of the business.

FURTHER READING

Arghern, A.A. (1997). Designing mature Internet business strategies: The ICDT model. *European Management Journal*, 15(4), 361-369.

Davidow, W. H. and Malone, M. S. (1992). *The Virtual Corporation: Structuring and Revitalizing the Corporation for the 21st Century*, New York, NY: Harper Business Press.

Goldman, S. L., Nagel, R. N. and Preiss, K. (1995). *Agile Competitors and Virtual Organizations*, New York, NY: Van Nostrad Reinhold.

Henderson J.C. and Venkatraman, N. (1993). Strategic alignment: Leveraging information technology for transforming organisations, *IBM Systems Journal*, 32(1), 4-16.

Moreton R. and Chester, M. (1997). *Transforming the Business: The IT Contribution*, London, UK: McGraw Hill.

Poon, S. and Swatman, P. M. C. (1997). Small business use of the Internet: Findings from Australian case studies. *International Marketing Review*, 14(5), 385-402.

Tetteh, E. O. and Burn, J. M. (2000). "SM" e-business-Strategies for success. In Khosrowpour, M. (Ed.), *Challenges of Information Technology Management in the 21ˢᵗ Century*, Hershey, PA: Idea Group Publishing, 406-409.

Venkatraman, N. (1994). IT-enabled business transformation: From automation to business scope redefinition. *Sloan Management Review*, Winter.

REFERENCES

Australian Bureau of Statistics. (1999). *Small and Medium Enterprises: Business Growth and Performance Survey 1997-98*, ABS Catalogue No. 8141.0, Canberra.

Boudreau, M., Loch, K. D., Robey, D. and Straud, D. (1998). Going global: Using information technology to advance the competitiveness of the virtual organisation. *Academy of Management Executive*, 12(4), 120-128.

Grenier, R. and Metes, G. (1995). *Going Virtual: Moving your Organisation into the 21ˢᵗ Century*, New Jersey: Prentice Hall Computer Books.

Rayport, J. F. and Sviokla, J. J. (1995). Exploiting the virtual value chain. *Harvard Business Review*, November-December.

Sieber, P. (1998). Organisational virtualness: The case of small IT firms. In Sieber, P. and Griese, J. (Eds.), *Organisational Virtualness, Proceedings of the VoNet-Workshop*, April 27-28, Bern, Simowa-Verlag.

Swartz, E. and Boaden, R. (1997). A methodology for researching the process of information management in small firms. *International Journal of Entrepreneurial Behaviour & Research*, 3(1), 53-65.

Tetteh, E.O. (1999). From business networks to virtual organisation: A strategic approach to business environment transformation in online small and medium-sized enterprises. *Proceedings of the 10th Australasian Conference on Information Systems (ACIS' 99)*, Wellington, New Zealand, December, 980-992.

Tetteh, E. O. and Burn, J. M. (2001). Global strategies for SMe-business: Applying the SMALL framework. *Journal of Logistics Information Management*, 14(1-2), United Kingdom: MCB University Press.

Venkatraman, N. and Henderson, J. C. (1998). Real strategies for virtual organising. *Sloan Management Review*, 4, 33-48.

BIOGRAPHICAL SKETCH

Emmanuel O. Tetteh is a Research Associate of the We-B Research Centre, School of MIS, Edith Cowan University, Western Australia. His current research interests include electronic business strategies, Internet infrastructures in SMEs, virtual organisations, information and communications technologies (ICT)-enabled business transformation, and national information infrastructures (NII) strategies and case research. Since 1998, he has designed and developed a number of online environments for research and teaching in e-business and virtual organisation. He has more than six years experience in ICT policy research at the Science and Technology Policy Research Institute (STEPRI) of the CSIR, Ghana. He has also worked as a Computer Systems Engineer for a leading computer systems vendor in Ghana. Mr. Tetteh has a number of refereed publications on e-business and Internet applications in SMEs.

Dancing with a Dragon: Snags in International Cooperation Between Two IT Companies

Yi Wei and Sirp J. De Boer
University of Twente, The Netherlands

EXECUTIVE SUMMARY

International strategic alliances are an increasingly popular way for companies to expand their operations beyond national boundaries. For small and medium-sized ICT enterprises, this route provides interesting characteristics. However, success does not come automatically in international strategic alliances.

This chapter applies the aspects of strategic fit, resource fit, cultural fit and organizational fit to analyze the cooperation between a small Dutch IT company and a major Chinese Internet content provider. Their cooperation does not run smoothly, and the chapter describes the background and presents details on their agreement. After both opposing views are described, the case is analyzed.

The chapter concludes that in this case study, there are several misfits. Even though the major misfits are in the strategic area, these could be resolved by improving the mutual understanding of positions and objectives. Moreover, it is more appropriate to approach the intended cooperation as a virtual alliance.

BACKGROUND

In today's global economy, increasingly companies are seeking partnerships to stay ahead of the competition or sometimes just to survive. They are fueled by a number of global developments such as internationalization of markets, increasing complexity of technologies, shortening of product life cycles, high economic uncertainty and increasing speed with which innovations take place (Faulkner, 1995; Douma, 1997). Especially the technology advances in ICT are permitting an increased flow of information across borders, and enable small and medium-sized enterprises (SMEs) to compete globally regardless of the physical locations (Castells, 1996; Ball & McCulloch, 1999). However, few firms have the human, financial and technical resources to 'go-it-alone' in every market and with every product, especially SMEs. A key benefit in forming alliances is clearly the ability to leverage the company's success into new customers and new markets (Gale, 1994). Therefore, global alliances form one of the quickest, less risky and cheapest ways to develop a global strategy, placing a fresh set of demands on companies, while it indeed has been identified as an important strategy option for small technology-based firms (Porter & Fuller, 1986; Forrest, 1990; Bairdl, 1997; Deresky, 1997).

In defining strategic alliances literature provides many terminologies such as alliances, cooperation, joint venture, coalition and collaborative agreement. However, there is no difference in the basic meaning among those definitions, which includes cooperation, a set of agreed-upon goals, shared resources and competence, and remained independence (Porter & Fuller, 1986; Yoshin & Rangan, 1995; Faulkner, 1995; Douma, 1997; Das & Teng, 1997; Gulati, 1998). Therefore, Mulyowahyudi (2001) concludes that the international strategic alliance is cooperation between companies from different countries that unite to pursue a set of agreed-upon goals through continuously sharing their respective complementary assets and core competencies. And each of them retains its independence and identity, to gain mutual benefit and to strengthen their competitive advantage.

The success of strategic relationships is driven by both structural elements and social elements (Van der Zee & Van Wijngaarden, 1999). However, starting and successfully continuing an alliance is more troublesome than many managers had expected. The high probability of failure in alliances is now common knowledge; the estimates of the success rate of cross-border alliances and acquisitions range from only about one-third to one-half (Rodrigues, 1996; Hoeckling, 1995). There are several principal causes of failure or difficulties in business partnerships such as compatibility of partners, corporate cultures, structure, ownership, controls, exits options, mismanaged expectations, access of the information, distribution of earnings, potential loss of autonomy and changing circumstances (Rodrigues, 1996; Griffin & Pustay, 1998).

Selecting an appropriate partner is the most important issue in forming a strategic alliance. In general, what a company looks for in a partner is the ability to contribute complementary strengths and resources to the alliance, which can help them to overcome any weaknesses that prevent or inhibit the ability to achieve its business objectives. De La Sierra (1995) states that there are three Cs that can measure a prospective partner, namely compatibility, capability and commitment. However, many scholars have found out that the key issue in partner selection is to assess the degree of possible fits to the other partner's profile. Fit is primarily about the question whether successful cooperation is possible, given the strategic background, objectives and organizational characteristics of the potential partners. Douma (1997) presents a fit model in which the strategic fit is the starting and indicating point, followed by cultural fit, operational fit, human fit and organizational fit, leading to an effective and successful alliance. Lasserre & Schutte (1995) quote that there are basically four different types of fit: strategic, resources, cultural and organizational. In this chapter, we apply the *strategic, resource, cultural* and *organizational* fits, and strategic fit as the major measure of the business cooperation.

Strategic fit is whether the partners' strategies, objectives or motives are mutually dependent and compatible, and the alliance is of strategic importance to the partners' competitive position. It is primarily concerned with the question whether there is a sufficient strategic basis for a successful cooperation. If the alliance does not have strategic importance for both partners, they will probably be insufficiently committed to making the necessary efforts and concessions for the alliance (Douma, 1997).

The general reason behind business cooperation is resources which firms need in common (Klofsten & Scharberg, 2000). *Resource fit* requires the partners to be willing and able to contribute to the critical resources, assets and competencies needed for the competitive success.

Cultural fit involves the individual corporate culture, industry culture, and national or ethnic culture, and these differences in culture influence three major aspects of management of particular relevance in the partnership: business objectives, competitive approaches and management approaches (Redding and Baldwin, 1991; Mead, 1994). The human fit is in essence about the mutual trust between the partners involved in the alliance (Douma, 1997).

Organizational fit is described as a situation in which the organizational differences do not hinder in the functioning of the alliance, and the partners have a shared vision on the alliance design. It is concerned with the question whether the alliance design the partners intend is effective, given the alliance objectives and possible organizational differences between the partners (Douma, 1997). The decision-making and control mechanisms used by partners should be conducted to good communication and effective monitoring in order to realize their objectives (Lasserre & Schutte, 1995; Faulkner, 1996).

Several authors have suggested, where both partners have equal strength, equal ownership, equal sense of commitment, mutual trust and flexible, same business vision and in core business area, effective communication, and strong and independent management, there is greater success in an alliance (Bleeke & Ernst, 1991; Hoeckling, 1995; De La Sierra, 1995). In the case of unequal parties, Faulkner (1995) states that an alliance between a very small and large partner is unlikely to be successful in the long run as the large one might acquire the smaller one or the alliance may break up due to the differing interests.

A small Dutch IT firm ('Monkey') aimed at global business expansion by forming a strategic alliance with a large Chinese enterprise ('Dragon'). Currently, the business partnership is standing at the deadlock situation just after the cooperation started one year (the latest message is both parties have agreed to stop the cooperation). This chapter applies the aspects of strategic fit, resource fit, cultural fit and organizational fit to review the whole process of their business cooperation and examines the major causes, which led the partnership to such a situation. The methods used for the case study include student assignments and interviews with the key persons in both parties. The main objective is to define whether an alliance's chance of success may be enhanced by better preparation by analyzing the failure alliance, which is significantly useful for theoretical and practical studies for reducing the cooperating risk.

The chapter describes the background and presents details on their agreement. After describing both opposing views, the case is analyzed along the four mentioned aspects of fit, and it concludes that there are several misfits within the cooperation. Even though the major misfits are in the strategic area, these could be resolved by improving the mutual understanding of positions and objectives. Moreover, it is more appropriate to approach the intended cooperation as a virtual alliance in the Internet age. From such a perspective it is imperative to develop trust in order to truly develop a successful strategic, i.e., long-term alliance.

SETTING THE STAGE

'Monkey' is a small IT company that was founded in 1995 by an entrepreneur who moved from the computer hardware field to the software sector. Currently, there are 15 employees working in its head-office located in Enschede, The Netherlands, for the European market. It has developed software for quickly creating a professional website page complete with picture and text, search-friendly databases, which it offers to companies and individuals for presentation of their products and services on the Internet. Even though 'Monkey' is small, some smart steps had been taken. For instance, it started cooperation with the University of Twente in 1999 to develop software for further innovations of computer networks, which could reduce its risk and lower capital requirements. From Monkey's point of view, Version I/Version II supplies the most innovative software available world-wide, and there is no product on the market today that competes directly with its products, while its products' simplicity and accessibility are expected to attract large numbers of customers from many business sectors. The current and potential customers of Version I/Version II in Holland (Dutch/English version) mainly consist of ISPs (Internet Services Providers), publishers, advertisers, manufacturers, trading houses and some sports clubs.

This colorful online method *Version II* is a software program based on five logical steps. In a simple and straightforward manner, it enables the user to place graphics, text and keywords on professional webpages in less than a few minutes, and the user or advertiser can create and update those pages at any time without having advanced computer or networking knowledge. *Version I* is a database (server software) used in combination with Version II. The software module automatically deals with all the pages and arranges them into internal search machines, websites, catalogues and portals. Once a webpage is ready for distribution, a simple push of a button sends the page with a number of keywords to a global Internet guide, and there is no limit to how often and how many pages can be changed or updated. The integrated interactive marketing methods, as for e-commerce, etc., can be linked to various payment systems. Moreover, apart from English, all other languages can be integrated (Tang, 1999).

'Monkey' is small, but aiming to go global, especially to the Far East. In 1998, the founder of 'Monkey' realized that China, which is one of the fastest developing economies and Internet markets, could be a huge potential market for Monkey's products. However, due to lack of the resources and experience, 'Monkey' cannot do it alone and it must find suitable partner(s) to develop the market together.

'Dragon' is a hi-tech shareholding enterprise located in Shenzhen, China, which specializes in ICP operation and related technical support. It started operation in October 1995, and the registered capital is US$23 million. By the end of 1999, its operations covered 108 cities and provinces in China, the total assets US$240 million, and the number of employees was 2,360. Their main services include ISP (Internet Services Provider), ICP (Internet Contents Provider), IPP (Internet Phone Provider), intranet and I-fax, etc. In 1999, it had 244,000 subscribers and the total income was US$26.3 million. The expected income for the coming years are year 2000 US$115 million, year 2001 US$464 million and year 2002 US$1.4 billion.

The business development periods of 'Dragon' as follows:
(1)   Preliminary period (1996-1997)
  • ISP services to individual subscribers
  • Commercial e-mail and information publication services to SMEs
(2)   Middle period (1998-present)
  • Trade commercial network platform services for government sub-scribers, and city information port platform services for local government
  • Wide intranet platform services for large-scale enterprises, and commercial information and electronic trading services for SMEs
  • ICP platform and network charging services for information services
  • Information services, professional services and network goods purchase for individual subscribers, etc.
  • I-fax: network fax
(3)   Long-term

With the further strengthening of the bandwidth of China Telecom and the loosening of telecommunication policies, the business to be opened in the future is as follows:
  • I-Phone: network telephone
  • I-Meeting: visual-information meeting
  • VPN analog professional network
  • Multimedia professional services

As mentioned before, 'Monkey' wants to enter China's market. After spending several months searching and negotiating, 'Monkey' and 'Dragon' had agreed to set up an alliance to market Version I/Version II products into Mainland China. The agreement—a cooperative venture between 'Monkey' and 'Dragon' dated November 24, 1999—specified that 'Monkey' would develop the new product (*Version III*) and transfer the technical knowledge for the new venture. 'Dragon' would provide capital for financing the operation equipment and office in Shenzhen (where the new venture locates), and is responsible for looking for the potential customers by using its current business network. Furthermore, 'Dragon' would recruit employees for the new venture. *Version III* is based on its product—Version I/Version II, but the original languages of the product are Dutch and English. To meet the requirements of Chinese customers, it should be changed to Chinese/English. Moreover, Version III is a product for Chinese customers who are different from Monkey's current customers in Europe. The potential customers for Version III include ISPs, publishers, government, wholesalers, industrial associations and big manufacturers. Chinese users can just go to the Version III website to create pages including keywords, text and pictures in a few minutes. They can choose pictures from their own sources or from Version III databases. Those pages are online and have host names, and will be automatically linked to 25 international search engines. In detail the requirements of Version III include:

Chinese and English: all pages have Chinese and English versions. Visitors around the world can choose Chinese or English to view the pages. Though Version III is written in simplified Chinese, it also can be used to make traditional Chinese pages that are also suitable for Taiwan and Hong Kong markets.

Technically easy for use: customers can make professional pages in a few minutes and every task can be done with only five color buttons on the Web. It is easy to learn and people without any PC and Internet knowledge can use it.

Multiple functions, and flexibility and interaction: the webpages created by Version III have completely automatic content management, automatic administration and control. Customers can make changes at any time and as often as they like at zero additional cost.

Cheaper price: by aiming at the huge potential customer base and the market condition, the price of Version III should be much cheaper than Version I/Version II in its home country/Europe.

CASE DESCRIPTION

The cooperation between 'Monkey' and 'Dragon' started in 1998, and the details of their cooperation follow:

The Initial Contact and Negotiation Period (1998-mid 1999)

At this stage, the contacts between both companies were based on the discussion of product/technology requirement, market condition and cooperation items including financial, staffs, structure and the responsibilities. They did not have any relation before they started contacting, and the result was that communication did not run smoothly. To overcome the cultural barrier, 'Monkey' employed a Chinese MBA graduate from a business school in Holland, to act as the China project manager. Since then, the communication between the two sides was much easier.

The Innovation/Improvement Period (mid 1999-end of 1999)

Based on their agreement that had been signed at the previous stage, the major task at this stage is development of the new product—Version III. 'Monkey' spent several months employing a number of programmers to complete it as soon as possible, and by the end of 1999, the job had been done.

The Operating Period (January 2000 until present)

Since the new product, Version III, had been developed—the operating process was on track. The operating center is set in Shenzhen where the 'Dragon' headquarters is located, and some staff was transferred from 'Dragon' to this venture. 'Monkey' hired another Chinese MBA graduate who also gained his degree in The Netherlands to be its representative to work with that team.

To better understand their cooperation, some important items of their 'cooperative-venture' agreement are listed below:

1. Both parties agree to set up and own an Internet application platform. The platform is based in China and open to the world business users who can make full use of above-mentioned platform and respective resources advantages to promote Monkey's products in commercial use. The technical solution for the Internet called Version III (based on Version II and Version II development) and the contents in the Chinese and English portals will be 50/50 ownership of the two parties for the Chinese market. They intend to set up a joint venture in Mainland China, after a trial period, working together as exclusive partners. All the investments and the achievements by both parties will be transferred to the joint venture.

2. 'Dragon' will provide operating premises, network systems and platform resources to the Version III project, maintain and administer Version III server in China daily. It agrees to use its business networks to promote and sell Monkey's products and services within the Territory, and actively promote the use of Monkey's products and services on its network and introduce the software to all potential users. 'Dragon' will use all methods such as advertisement on different media and attending exhibitions to do the marketing and promotion, and 'Monkey' will do such outside the territory, such as the promotion on the Internet websites, the extra hosting on their servers and the promotion in all kinds of publicity. 'Dragon' commits and pays for sales and technical staff necessary to achieve the agreed sales target and makes the marketing plan for Version III and implements it with the approval of 'Monkey.'. It will promote and market Version II and its related products to the Chinese market.

3.   'Monkey' will provide English and Chinese versions of the Version III and its client software (Version I), and will supply and install a licensed copy of the Version II Web server suite of software on the hardware supplied by 'Dragon.' Parties will not charge each other for this activity. It will prepare mirroring servers for Version III outside China, and provide research and development to Version III and upgrade the software to the latest version. 'Monkey' will provide marketing instructions and master copies of its brochure and video to the 'Dragon' without cost, as well as technical online support, and provide technical support and training to 'Dragon' technicians. To meet the demand of the Chinese market, 'Monkey' will provide the administrative module of Version III for 'Dragon' to change interface, back-office administration, editing templates and so on, and ensure to update in time the Version III servers both in and outside China.

4.   'Dragon' and 'Monkey' intend to sell the Version III FREE E-commerce package. This package includes the homepage, business profile page, map, response page and five product pages. Version III includes the website, the catalogue, the e-commerce shop and also the presence in the Version III portal. 'Dragon' will have 70% of the e-commerce transaction fees and the revenue of the advertising, and 'Monkey' gains 30%. The estimate is to sell 10,000 packages in the first fiscal year, starting on the January 1, 2000. The target is 50,000 packages for the first year, and this figure will be adjusted according to the results in the Chinese market. Within 14 days after the end of each month 'Dragon' will submit a written report (sales declaration) to 'Monkey,' which details all the numbers of customers, and the total number and value of pages sold during the past month. Within 30 days of each sales declaration, 'Dragon' will submit a payment to 'Monkey' by telegraphic transfer, which is equal to 50% of the invoiced value of sales declared less, the costs of local taxes and byte account fees.

5.   'Monkey' has the right to check the administration of 'Dragon' by an independent financial controller, and the inspection is limited to the administration of 'Monkey' products and services. 'Monkey' also has the right to end this agreement before the period of five years is over if 'Dragon' does not achieve the minimum annual turnover expectation.

At the beginning, they scheduled to create an Equity Joint Venture, but afterwards, they started as a cooperative venture with no financial investment involved. However, based on their contract, the cooperation is more likely a Joint Marketing Alliance, or 'Dragon' is acting as Monkey's China market main agent/distributor, because:

1. They have not set up an independent company/work team for the venture. The staff who work for this project belong to a department/ office of 'Dragon' and 'Monkey' only have one representative person as the bridge to work with them. The 'Monkey' project is just one of the tasks of most of the staff involved.
2. No financial departments from either party have been involved in this project (according to an interview with 'Dragon'' managers), and the venture does not have a separate bank account. The financial relation between the two parties is simple, that of only one-way flow: after the product had been sold and 'Dragon' received the payment from customers, 'Dragon' would transfer the profit in US$ based on the agreement to Monkey's account in The Netherlands.

Current Challenges/Problems Facing the Organization
Opposing Views

Now that both partners have been operating for almost one year, there is a deadlock situation in the cooperation between the two parties. Based on several interviews with both parties, the problems and complaints can be summarized, as below.

From Monkey's point of view, the main problems of the unhappy marriage with 'Dragon' are:

1. 'Dragon' is financially and technically weak in promoting Version III, while 'Monkey' is unwilling to input more investment since every small and private company has the same problem—capital shortage and hesitation in investing in such high-risk areas unless it sees certain returns.
2. The management team of 'Dragon' is poor in terms of leadership and employees' quality and loyalty, and also lacks proper understanding of the product and market.
3. The geographical distance makes the communication between the two parties weak, and the representative of 'Monkey' cannot deal with the matter due to some unknown factors. Therefore, 'Monkey' has no adequate control in the business.
4. Still related to the capital shortage, it cannot finance an independent working team that they do need in China without having gained profit from the business.
5. The cooperation period has been set for five years, and 'Dragon' acts as Monkey's only agent in China. A classic error for such cooperation is that nothing guarantees that the distributor will possess the corporate culture appropriate to a slow cash-flow cycle, or that it will not simply

bring a trading mentality to a few rapid cash-flow deals instead of developing a long-term approach to the product.

6. 'Dragon' has offices in more than 100 cities around China, but it only actively engages its office in Shenzhen (Guangdong), which means 'Dragon' is just targeting the product at the Guangdong province rather than nationwide. However, Beijing and Shanghai as the two major cities/regions count almost half of the online users and websites in China. Therefore, the scope of the current marketing strategy is too narrow. And the restrictions in their cooperation contract do not allow 'Monkey' to work with other parties in China, which also endangers its products' competitiveness.

'Dragon' blames 'Monkey' on the following points:

1. The communication with 'Monkey' is very difficult even though 'Monkey' had hired a contact person in China and there is another Chinese MBA graduate in Monkey's head-office in Holland. For instance, one manager in 'Dragon' questioned why 'Monkey' always communicates with him via email in English while 'Monkey' does have a Chinese employee; this matter makes him respond in English as well but surely not clearly and understandably.

2. Within 'Dragon' there is not much money available to market Version III product and 'Monkey' refused to invest as well. Therefore, the budget for advertisement is limited and few people and potential customers in China know the product. In addition, the e-market in China is still small and the transportation system is still poor, and it requires hard work and effort. It is hard for 'Dragon' to bring the Version III product alone to their potential clients aside from their major business activity. However, 'Monkey' cannot understand and accept the fact.

3. The target customer group for the Version III project is SMEs since big companies already have the 'electronic' people to make their own websites but SMEs lack capital and experience to do so. Yet, 'Monkey' is always aiming at the large firms.

4. The Version III product only makes use of Simplified Chinese and English although 'Monkey' had promised that the product could be used under the Traditional Chinese system as well. It suits the Mainland China market, but it cannot be applied to other areas like Hong Kong and Taiwan, which are using the Traditional Chinese system, narrowing the market scope. However, the investment from those areas account for more than two-thirds of the total foreign direct investment in China, and most of those enterprises are export-oriented (which is the Version III

product target). Meeting the customer demand is very important and such a function should be added. Moreover, 'Monkey' does not have the up-to-date knowledge of the Chinese market and its judgment is based on the European market and always with high expectations.

5. For protection purposes, 'Monkey' does not provide the program source codes to 'Dragon' and did some training for 'Dragon'' technicians. There have been some technological problems that some customers had met. However, 'Dragon' did not get the technical support from 'Monkey' because 'Monkey' insisted on seeing the payment/return first. Therefore, the demand/requirement of the users cannot be met quickly. As one manager in 'Dragon' states: "Believe us, give me the program of Version III, I can resolve the problems at once. But now, we have to send emails to 'Monkey' in Holland, the time will pass and the time is important for getting the customer and money in the market."

6. 'Dragon' sees the IP Fax and IP Phone as its priorities for the business activities, giving the opportunity to 'touch' their clients monthly and closely when they use the services. They are willing/had started to add Version III to this product package and currently a group of 1,000 IP Fax/Phone clients already received the Version III product. 'Monkey' will get a certain share from the income of the package sales. However, 'Monkey' hesitates about such an idea and never responded to the 'Dragon' plan.

Case Analysis

Since the cooperation between 'Monkey' and 'Dragon' currently is at the deadlock situation, it is appropriate to review the whole process. Therefore, the different fits between the two parties are used to evaluate their cooperation.

Strategic Fit

In strategic alliances, firms may have different strategic positions in terms of the strategic importance of that particular business (core business or peripheral) and how it fits with the overall portfolio of the partners (Lorange & Roos, 1992). The main question the partners must pose both themselves and each other is whether there is sufficient strategic fit to justify the alliance. Individual interests are weighed against the anticipated advantages of the alliance, and the potential risks linked to it. It is often assumed that a successful alliance is only possible when the partners have similar objectives. Overlap in objectives is unavoidable, and there must be a common strategic interest to motivate cooperation. A major difference in strategic fit diminishes the stability of the cooperation, in that the partners' commitment to the alliance will correspondingly differ (Douma, 1997).

In this case, on the one hand, 'Dragon' is one of the largest Chinese Internet companies and it is part of a large conglomerate of companies operating internationally, whose activities among other things involve electricity supply, telecommunication, Internet, networks and trade. It is aiming to become a powerful ISP player in China, and the main business objective for it is similar to all Internet companies—getting more users and then go public. However, most of China's ISPs are having a hard time due to the high fees, still small market and the competition, and they have to extend their business activities from providing basic services/contents to other added-value services (Wei, 2000). Therefore, 'Dragon' sees the IP Fax and IP Phone services as its priorities for strengthening its position in China's market. In addition, based on the interviews with the managers in 'Dragon,' this company aims to enter the Western market as well and they hope its business partner can be an agent or bridge for 'Dragon' and its customers. On the other hand, 'Monkey' is a small but innovative software development company. Like the other small high-tech firms, it also wants to grow rapidly. The booming Chinese Internet market has provided the huge opportunities for 'Monkey' to achieve its dream. However, as it lacks the necessary resources and experience needed to gain the market share, it has to find a strong and suitable partner who can share the risk and provide efforts with it.

Nowadays, the lifecycle of IT products is becoming shorter and shorter. The product Version III is one of Monkey's core competencies. A manager had pointed out during the interview that the Version III product is about three or four months ahead of the competing software products in China. Since time for 'Monkey' is so important (especially upgrading the products in time), the success and quick market development and expansion in China is its business strategic priority. However, 'Dragon' sees Version III as just one of the added services to its product package for the end users, and the marketing plan of the project should be based on its own business strategy. For instance, 'Monkey' expected to start the project nationwide in China through the 'Dragon' whole business network, but 'Dragon' insisted on starting the project from its five offices in Guangdong province first, and then, after two years, expansion of the business activities will be made to other regions such as Jiangsu province, Shanghai and Beijing. In addition, 'Dragon' does not recognize that 'Monkey' can be a major agent/bridge for its business and customers to enter the European market. Therefore, 'Dragon' will not commit the expected efforts and resources to the project besides its own objectives. The different viewpoints from the beginning have caused the different expectations and many misunderstandings, and it has led this cooperation to an unhappy result.

Resource Fit

The resource matching also causes a large gap between the cooperation parties. In strategic alliances, firms may come to share complementary resources, and they may have different views as to what resources are and their potential value. Especially in cooperation between large and small firms, larger firms may find that smaller firms lack everything they promise and it is particularly difficult for large firms to admit that they have something to learn from a smaller partner. Meanwhile, small firms may blame large firms for being distracted by commitments to dominate markets (Hamel, 1991). For the smaller one, it may take advantage in flexibility and innovation, but weak in liabilities of smallness and newness. However, the smaller one can overcome inherent disadvantages by leveraging its resource base; that is, the firm seeks to do more with less.

In this case, on the one hand, 'Monkey' as a small private company is strong in technology but too weak in finances and experience in China. The weakness has resulted in that it cannot afford a working-team (management and technical) in China individually and also cannot invest more money for the promotion and marketing of its product. Therefore, it only can provide the technical base and the limited resources (people and money) to the new venture, while hoping that it can get the maximum market share and return in China's market as quickly as possible to strengthen its position on the negotiation table with venture capitalists. However, all hopes depend on its partner's whole business network, and the efforts and speed for marketing the Version III product.

On the other hand, 'Dragon' is a nationwide group of companies with 80 licenses, more than 100 offices and 2,000 employees in China. However, the major business activities of 'Dragon' are focused on Guangdong province, and most of its customers are from this region. The offices in other regions are mainly part of the group's long-term business strategy and many of them are cooperation units with other parties; therefore, the relation between branch and branch, or even branch and the head-office is not as close as 'Monkey' had expected. Also, in some points, 'Dragon' is big, but like most ISPs, it is also eager to attract more external financial sources to promote its new product package. The disappointed capital markets around the world, especially in the technological field this year, have resulted in negative impacts in most high-tech companies' business strategies and activities, which means they would not do anything further without seeing the return clearly. Meanwhile, 'Dragon' hopes 'Monkey' can send permanent marketing and technical staff to China or give the source codes of the software to its technicians for quickly responding to customers' requirements and further development, but 'Monkey' refused to do so.

As mentioned previously, 'Dragon' did not fully use its business network to market the Version III product into the whole China market but just focused on Guangdong province. In addition, both parties are expecting the other side could provide the necessary resources (people and money) to promote Version III product but neither is willing to do so. Therefore, the limited budget for the Version III project has resulted in a difficult and slow marketing process.

Cultural Fit

The cultural differences or communication problem between parties may create problems in their business cooperation. Especially in the cooperation between IT firms, one party may have very specialized know-how which cannot be readily transferred to the other for further development. The cultural variables in information systems and context surely underlie the many differences in communication style between Dutch and Chinese. Cultural fit involves the individual corporate culture, industry culture and national or ethnic culture (Redding and Baldwin, 1991; Mead, 1994). However, our analysis will be based on the trust and language areas.

Trust is the expectation shared by the parties involved that they intend to meet their commitments to each other. The development of trust-based relationship is thus a key issue in inter-organizational exchanges in the network era. Distrust arises from communication problems, personal differences, lack of internal support, conflicts caused by implementation of human resource and technology transfer policies, changing goals, and strategic mismatch of long-term interests.

As 'Monkey' is a Dutch-based company and 'Dragon' is a Chinese firm (and even though 'Monkey' had employed two Chinese staff), the language difficulty is still remaining within the business cooperation. For example, the daily communication language they are using is still English through emails and faxes. As a key person in 'Dragon' stated, "It is difficult for me to write and read emails in English, the only thing I can do is to reply emails in very simple English. Therefore, surely I cannot present my points clearly and it causes much confusion and misunderstanding." And he also thinks that the Chinese person in the 'Monkey' head-office in Holland cannot use the Chinese software system and, therefore, the communication language between the two Chinese is English.

The trust-based relationship has not yet developed between the two parties, especially from the 'Monkey' side. It is common that the smaller party is in the weaker position when a cooperation is formed between small and big companies. Therefore, the smaller one always uses the defensive strategy to protect its right and position. This issue can be illustrated as follows. Once a company bought Version III but had not paid for it one month after purchasing; then they could not log in anymore because 'Monkey' only gave them a temporary account, which could be used for one month. Later the company paid to 'Dragon' but 'Monkey' still refused to reopen the account for the client as 'Dragon' had not transferred the fee according to their cooperation agreement to Monkey's account yet. In such a case, the trust between the two parties is nearly zero.

Organizational Fit

In alliance design, management constraints may be the most significant difference between small and large firms. Small firms were mainly behavioral, i.e., management, dynamism, organization and flexibility, whereas large firms were mainly material (Rothwell, 1991). Four factors were identified to determine the degree of organizational fit or not. These are flexibility, management control, complexity and trust (Douma, 1997). Flexibility is one of the most important advantages in strategic alliance to compare to a merger or acquisition. This means that not all the company's resources need to be dedicated to one strategic option, and that a strategic alliance may in principle be terminated, if it should turn out not to yield the results expected. Management control is primarily concerned with the influence the individual partners have on the alliance policy and activities, and with the opportunity this gives to check whether the partner is fulfilling its obligations. More control for one is at the expense of the other's influence. Finding the right balance is essential, and demands continuous attention from the partners. The relative bargaining power is determined by the partners' strategic positions (need, dependency), and the resources (money, people, know how, etc.) the partners are prepared to commit to the alliance. According to Killing (1988), alliance complexity involves the question whether or not the alliance design enables effective management control. However, here we discuss the trust again. Rational trust means that the partners assume that the opposite party has such an interest in the alliance's success, that it will not display opportunistic behavior and will meet its obligations. Personal relationships and informal contacts between the managers seem to play a role, for there is generally a lower barrier to talk on cooperation here.

The alliance design of 'Monkey' and 'Dragon' is simple in that 'Dragon' takes most responsibility in providing the marketing and technical staff and 'Monkey' trains those staff in more details of the product. Therefore, 'Dragon' takes a strong bargaining position compared with 'Monkey.' However, the bargaining position imbalance in the project is due to inequality and the limited strategic fit between the two parties. In such a situation, there is a very great chance the alliance will only yield advantages for one of the two partners, which will endanger its stability. Although a contact person of 'Monkey' was sent to join the work team, the contact person does not have the power for decision making. Every step and detail are under discussion between 'Dragon' in China and 'Monkey' in Holland via fax, email and telephone. From this viewpoint, it is more like a virtual cooperation (see Davidow & Malone, 1992; Lucas, 1995). A virtual organization is an electronically networked organization that transcends conventional organizational boundaries, with linkages which may exist both within and between organizations, to create new markets, offer new products/services (value added) or assure flexibility in responding to new market requirements (Grenier & Metes, 1995; Pletsch, 1998; Burn et al., 1999). Communication is fundamental to any form of organizing, but it is especially important in virtual organizations (Mowshowitz, 1997; DeSanctic & Monge, 1998), because relationships in a virtual environment appear more difficult as individuals work in different physical contexts, and the shared understanding of information and communication patterns is less easily achieved (Holland, 1998). While electronic communication is heavily influenced by surrounding social norms, Handy (1995) questions whether one can even function effectively in the absence of frequent face-to-face interaction in a virtual environment. Therefore, the trust-based relationship is even more important when enterprises are moving toward the virtual cooperating way, while the benefits of trust tend to be long term, whereas benefits of acting untrustfully tend to be short term (Jarvenpaa & Shaw, 1998). During the interviews, both parties complained about the slow response from the other side and this matter has even appeared in serving customers.

Discussion and Conclusion

Klofsten and Scharberg (*2000*) state that the barriers between small and large IT firms can be divided into three types: differential interests, resources and size; differential assets, secrecy and trust; and cultural differences and communication problems. From the above analysis, those barriers all can be found in the 'Monkey' and 'Dragon' case.

In the first sense, the sizes unbalance and the different strategic importance of their Version III project have shown there is no sufficient *strategic fit* between the two parties, even though both parties have provided their core competencies— Version III product and business network in China. However, since 'Dragon' sees Version III as just one additional product besides its own services, it will not commit its whole business network and resources to support this project.

Although China is one of the fastest growing nations in the World in Internet development, its market conditions are still underdeveloped. Most potential customers lack IT experience and knowledge; promoting the Version III product in China's market successfully requires large amounts of investment in advertisement and even in pre-education of end users. At this point, 'Monkey' is too small in size and too weak in its own financial sources, while 'Dragon' is not willing to provide the necessary investment either unless the project belongs to its own business priority and the partner provides the same amount as it does. Therefore, the resource gap leads the cooperation to running difficult and slowly.

'Monkey' had prepared well for overcoming the cultural difference with its Chinese partner. To do so, it had employed two Chinese MBA graduates who gained their degrees in Holland and had strong knowledge in understanding the Dutch and Chinese cultural difference, and located one in China and one in its head-office in Holland. However, in practice, the cultural gap and communication problem still remains, and it results in a total lack of trust between the two partners.

The alliance design in this case is simple. Due to the low strategic fit between the two parties, the bargaining position imbalance is quite distinct: 'Dragon' takes more power and control in the alliance. Such a situation will endanger the cooperation stability. In addition, since the daily communication between the two partners is via electronic form such as email, fax and telephone, the cooperation form is more like the currently widely discussed virtual organization (Alliance) form. In such a form, the base or the heart is trust. The benefits of a trust-based relationship tend to be long term, whereas benefits of acting untrustfully tend to be short term.

Neither 'Monkey' nor 'Dragon' had expected the alliance to stand at such a situation. Since they are considering ending the cooperation or changing the cooperation form, they should take into account the fits (strategic, resource, cultural and organizational) no matter whether they are looking for new partners or restructuring the cooperation. However, some authors have presented some suggestions to overcome those barriers. Different firm sizes may create differential interests in cooperation, but they may also lead to higher profits if the larger partner takes the lead (Killing, 1983). Nevertheless, the resource imbalances need not create differential interests in cooperation if the partners had defined similar interests and objectives when forming the alliance. The cultural differences can be overcome through cultural complementarily. In the long term partners must understand cultural differences between each other, have a will to compromise, have mutual trust and share a strong commitment to project goals (Klofsten & Scharberg, 2000). Building up effective alliance design and trust-based relationship is the long-term benefit for cooperation partners. In addition, developing conflict resolution techniques is an important step that can help resolve problems while they are small and still solvable (Klofsten & Scharberg, 2000).

ACKNOWLEDGMENTS

We would like to thank Mr. Johan De Kool and Ms. Claudia Stijlen for their valuable interviews and analysis in China, and the contribution of most background materials and suggestions from Mr. Liujin Tang. We express our appreciation to the managers in both companies who have provided the opportunities for interviews.

We have disguised the names of the companies involved to protect their interests, while at the same time making the interesting material available to a wide public. The chapter title alludes to the challenge of doing business with and in China.

FURTHER READING

Ball and McCulloch. (1999). *International Business—The Challenge of Global Competition*, 7th edition. McGraw-Hill Co., USA.

Burns, R. (1998). *Doing Business in Asia: A Cultural Perspective*, Addison Wesley Longman Australia Pty Limited.

Davidow, W. H. and Malone, M. S. (1992). *The Virtual Corporation,* New York: Edward Burlingame Books/HarperBusiness.

Douma, M. U. (1997). *Strategic Alliances: Fit or Failure*, Ph.D. Thesis, University of Twente, The Netherlands.

Faulkner, D. (1995). *International Strategic Alliances: Cooperating to Compete*, London: McGraw-Hill Book Company.

Gulati, R. (1998). Alliances and networks. *Strategic management Journal*, 19, 293-317.

Lucas, H. C. (1995). *The T-Form Organization: Using Technology to Design Organizations for the 21st century,* San Francisco, CA: Jossey-Bass Publisher.

Porter, M. E. and Fuller, M. B. (1986). Coalitions and global strategy. In Porter, M. E. (Ed.). *Competition in Global Industries*, 315-344, Boston MA: Harvard Business School Press.

Yoshino, M. Y. and Rangan, U. S. (1995). *Strategic Alliances: An entrepreneurial approach to globalization,* Boston MA: Harvard Business School Press.

REFERENCES

Baird, I.S. et al. (1997). The choice of international strategies by small businesses. In *Strategic Management in the Global Economy*, 3rd edition, John Wiley & Sons Inc, 175-184.

Ball and McCulloch. (1999). *International Business—The Challenge of Global Competition*, 7th edition, McGraw-Hill Co., USA.

Bleeke, J., Ernst, D. and McKinsey & Co. (1991). The Way to Win in Cross-Border Alliances. *Harvard Business review*, November-December.

Burn, J., Marshall, P. and Wild, M. (1999). Managing change in the virtual organization. *7th European Conference on Information Systems Proceedings*, 1, 41-53.

Burns, R. (1998). *Doing Business in Asia: A Cultural Perspective*, Addison Wesley Longman Australia Pty Limited.

Castells, M. (1996). *The Rise of Network Society*, Oxford, UK: Blackwell Publishers.

Das, T. K. and Teng, B. S. (1997). Sustaining strategic alliances: Options and guidelines. *Journal of General Management*, 22(4), 49-64.

Davidow, W. H. and Malone, M. S. (1992). *The Virtual Corporation*, New York: Edward Burlingame Books/HarperBusiness.

De La Sierra, M. C. (1995). *Managing Global Alliances: Key Steps for Successful Collaboration*, Addison-Wesley Publishing Company.

Deresky, H. (1997). *International Management—Managing Across Borders and Cultures*, 2nd Edition, Addison-Wesley Educational Publisher, Inc.

DeSanctis, G. and Monge, P. (1998). Communication processes for virtual organizations. *Journal of Computer Mediated Communication (JCMC)*, June.

Douma, M. U. (1997). *Strategic Alliances: Fit or Failure*, Ph.D. Thesis, University of Twente, The Netherlands.

Faulkner, D. (1995). *International Strategic Alliances: Cooperating to Compete*, London: McGraw-Hill Book Company.

Forrest J. E. (1990). Strategic alliances and the small technology-based firm. *Journal of Small Business Management*, July, 37-45.

Gale, T. P. (1994). *Integrated Supply & Alliances: Where is it taking industrial distribution?* November, Modern Distribution Management Journal.

Grenier, R. and Metes, G. (1995). *Going Virtual*, Upper Saddle River, NJ: Prentice Hall.

Griffin, R. W. and Pustay, M. W. (1998). *International Business: A Managerial Perspective,* second edition, Addison Wesley Longman, Inc.

Gulati, R., (1998). Alliances and networks. *Strategic Management Journal*, 19, 293-317.

Hamel, G. (1991). Competition for competence and interpartner learning within international strategic alliances. *Strategic Management Journal*, 12, 83-112.

Handy, C. (1995). Trust and the virtual organization. *Harvard Business Review*, 73(3), 40-50.

Hoeckling, L. (1995). *Managing Cultural Differences: Strategies for Competitive Advantage,* Addison-Wesley Publishers Ltd.

Holland, C. P. (1998). The importance of trust and business relationships in the formation of virtual organizations. In Sieber, P. and Griese, J. (Eds.), *Organization Virtualness, Proceedings of the VoNet-Workshop*. April 27-28.

Jarvenpaa. S. L. and Shaw. T. R (1998). Global virtual teams: Integrating models of trust. In Sieber, P. and Griese J. (Eds.), *Organizational Virtualness: Proceedings of the VoNet-Workshop*, April 27-28, 35-51.

Killing, J. P. (1983). *Strategies for Joint Venture Success*, New York: Praeger.

Killing, J. P. (1988). Understanding alliances: the role of task and organizational complexity. In Contractor. F. J. and Lorange. P. (Eds.), *Cooperative Strategies in International Business*, Lexington: Lexington Books.

Klofsten, M. and Scharberg, C. (2000). Barriers in cooperation between small and large technology based firms: A Swedish case study. *Proceeding for the 8th High-Tech Small Firms Conference*, Enschede, The Netherlands, May 22-23, 139-156.

Lasserre. P. and Schutte, H. (1995). *Strategies for Asia pacific,* MACMILLAN Press LTD.

Lorange, P. and Roos, J. (1992). *Strategic Alliances—Formation, Implementation and Evolution,* Cambridge, MA: Blackwell Publishers.

Lucas, H. C. (1995). *The T-Form Organization: Using Technology to Design Organizations for the 21st Century,* San Francisco, CA: Jossey-Bass Publishers.

Mead, R. (1994). *International Management: Cross-Cultural Dimensions*, Blackwell Publishers, Inc.

Mowshowitz, A. (1997). Virtual organization. *Communications of the ACM*, 40(9), 30-37.

Mulyowahyudi, A. (2001). *Success in Managing International Strategic Alliances*. Ph.D. thesis, University of Twente. The Netherlands (forthcoming).

Pletsch, A. (1998). Organization virtualness in business and legal reality. In Sieber, P. and Griese, J. (Eds.), *Organizational Virtualness, Proceedings of the VoNet-Workshop*, April 27-28, 85-92.

Pletsch, R. (1998). A framework for virtual organizing. In Sieber, P. and Griese, J. (Eds.), *Proceeding of the VoNet-Workshop, Orgnizational Virtualness*, April 27-28.

Porter, M. E. and Fuller, M. B. (1986). Coalitions and global strategy. In Porter, M. E. (Ed.), *Competition in Global Industries*, 315-344, Boston, MA: Harvard Business School Press.

Redding S. G. and Baldwin E. (1991). *Managers for Asia/Pacific: Recruitment and Development Strategies*. Hong Kong: Business International.

Rodrigues, C. (1996). *International Management: A Cultural Approach*, West Publishing Company.

Rothwell, R. (1991). External networking and innovation in small and medium-size manufacturing firms in Europe, *Technovation*, 11(2), 90-118.

Tang, L. J. (1999). *Create Successful Dutch-Chinese Equity Joint Venture*, Graduation Thesis, TSM Business School, The Netherlands.

Tayed, M. H. (2000). *International Business: Theories, Policies and Practices*, Pearson Education Limited.

Van der Zee, H. and Van Wijngaarden, P. (1999). *Strategic Sourcing and Partnerships: Challenging Scenarios for IT Alliances in the Network Era*, Addison-Wesley.

Wei, Y. (2000). *E-commerce in China: Developments and Issues*, Working paper in TDG, University Twente.

Yoshino, M. Y. and Rangan, U. S. (1995). *Strategic Alliances: An Entrepreneurial Approach to Globalization*, Boston, MA: Harvard Business School Press.

BIOGRAPHICAL SKETCH

Yi Wei is from the Faculty of Technology & Management, University of Twente, The Netherlands. His research focuses on the interrelationship between interorganizational systems (electronic commerce, EDI, electronic markets), new organizational forms, focusing on the business cooperation between China and the Western World.

Sirp J. De Boer is an Associate Professor in International Management in the Faculty of Technology & Management, University of Twente, The Netherlands. His practical experience includes a total of eight years of involvement in industrial development projects in 18 industrializing countries. His research focus is management of international technology transfer, while his geographic focus is China.

Section II

Regional Teaching Cases

Success in Business-to-Business E-Commerce: Cisco New Zealand's Experience

Pauline Ratnasingam
University of Vermont, USA

EXECUTIVE SUMMARY

The growth of business-to-business e-commerce has highlighted the importance of computer and communications technologies and trading partner trust for the development and maintenance of business relationships. Cisco Systems Incorporation, an international company, is now the second largest company in the world, behind Microsoft. Its solid financial performance is partly due to its early focus on the Internet as a channel to cut administrative costs, and boost customer service satisfaction. Cisco International provides end-to-end networking solutions which customers use to build a unified information infrastructure of their own, or to connect to someone else's network. The end-to-end networking solutions provide a common architecture that delivers consistent network services to all users (Cisco Fact Sheet, 2000). Cisco network solutions connect people, computing devices and computer networks, allowing trading partners to access or transfer information without regard to differences in time, place or type of computer systems. By using networked applications over the Internet on its own internal networks, Cisco globally is gaining contributions of at least NZ$825 million a year in operating cost savings and revenue enhancements (Cisco Newsroom, 2001). Cisco is today the world's largest Internet commerce site and sees financial benefits of nearly US$1.4 billion a year, while improving customer/partner satisfaction and gaining a competitive advantage in areas such as customer support, product ordering and delivery times (Cisco Fact Sheet, 2000).

Cisco International serves customers in three large markets, namely:

1. Enterprises including large organizations with complex networking needs, usually spanning multiple locations and types of computer systems. Thus enterprise customers include corporations, government agencies, utilities and educational institutions.
2. Service providers include companies that provide information services, including telecommunication carriers, Internet Service Providers, cable companies and wireless communication providers.
3. Commercial companies with a need for data networks of their own, as well as connection to the Internet and/or to business partners.

Cisco International (Cisco's headquarters) in San Jose, California, USA, has well over 225 sales and support offices in 75 countries. Cisco International wants New Zealand businesses to embrace the Internet and use it to be more efficient. The company worked with the NZ government on its e-commerce implementation plans at the summit held in late 2000. One of the aims of this forum was to encourage small and medium enterprises (SMEs in NZ) to go online.

Cisco NZ receives direction from its headquarters in San Jose, which monitors a global networked business model. A global networked business model includes an enterprise, of any size, that strategically uses information and communications to build networks of strong, interactive relationships with all its key constituencies. The global networked business model leverages the network for competitive advantage by opening up corporate information to all key-trading partners and employs a self-help model of information access, which is more efficient and responsive than the traditional model. The traditional model consists of few information gatekeepers dispensing data as they see fit.

The global networked business model is based on three core assumptions:

1. The relationships an organization maintains with its key constituencies can be as much of a competitive differentiator as its core products or services.
2. The manner in which a company shares information and systems is a critical element in the strength of its relationships.
3. Being 'connected' is no longer adequate. Business relationships and communications that support them must exist in a 'networked' fabric. Hence, by simplifying network infrastructures and deploying a unifying software fabric that supports end-to-end network services, organizations are learning how to automate the fundamental ways they work together.

Cisco NZ claims that the success of e-commerce depends on well-planned partnerships, mutual goals and trust. Cisco NZ's philosophy is to listen to their trading partners' requests, monitor all technological alternatives and provide customers with a range of options from which to choose. Thus, Cisco's experience in e-business has set the standard for e-commerce transformation and creating Internet solutions. This teaching case focuses on Cisco's experience with their trading partner, Compaq NZ, and the findings contribute to strategies on how businesses can succeed in e-commerce participation.

SETTING THE STAGE

Cisco Internationally operates in one of the most profitable commerce sites in the world with $7 billion transactions annually, approximately 20% of total global e-commerce revenues. Its Internet commerce applications have yielded more than $30 million annually in cost savings to the company (Cisco Fact Sheet, 2000).

Lack of the required skills, expertise and knowledge of the full potential of e-commerce applications has created a situation of lack of trust. Consequently barriers to participation in e-commerce activities arise due to uncertainties inherent in the current e-commerce environment. These, uncertainties, in turn, create a perception of increased risk thereby inhibiting the tendency to participate in e-commerce particularly from top management. Uncertainties reduce confidence both in the reliability of business-to-business transactions transmitted electronically and, more importantly, in the trading parties themselves (Hart & Saunders, 1997). Furthermore, the increased complexity of today's networking equipment requires trading partners to focus and develop an expertise around a specific technology (Cisco Newsroom, 2001). Despite the assurances of technological security mechanisms, businesses in New Zealand perceive that e-commerce transactions may be both insecure and unreliable. Similarly, preliminary research suggests that a perceived lack of trust in e-commerce transactions sent by trading parties using the Internet could be a possible reason for this slow adoption rate (Keen, 1999).

Cisco NZ claims that many of these companies are not using the Internet to its potential as they are frightened by what it means, and the potential costs that are involved. Hence, Cisco International (that is Cisco's headquarters in San Jose, California, USA) used relevant New Zealand case studies to show how businesses can use the Internet for competitive advantage. For example, ASB Bank e-solutions—a virtual joint venture between Telecom, EDS and

Microsoft, and Xtra—present a set of case studies as part of the program (Info-Tech, 2000).

In 1996, Cisco International embarked on an ambitious campaign to bring its largest customers into fold, and in the process introduced Cisco Connection Online (CCO), which is part of the global networked business model to Cisco in New Zealand. One of CCO's biggest accomplishments may be the way it has taken Cisco NZ trading partners. Employees within Cisco NZ have access to information and tools that allow them to do their jobs more proficiently, and prospects have ready access to information that aids in purchasing decisions. Trading partners have ready access to a variety of information and interactive applications that help them sell more effectively. Cisco NZ's gold trading partner Compaq NZ would rather have an automated way to tie their legacy purchasing or sales automation systems to Cisco NZ than to have their purchasing agents use Compaq NZ's website manually.

Hence, the global networked business is an open, collaborative environment that transcends the traditional barriers to business relationships and between geographies, allowing diverse constituents to access information, resources and services in ways that work best for them. Hence, Cisco International is not only the worldwide leader in networking, having supplied over 80% of the Internet backbone equipment, but is also a leading example of global networked business, leveraging its IT and network investments. In addition integrating them with core business systems, operational information to better support its prospects, customers, partners, suppliers and employees (Cisco Fact Sheet, 2000).

CASE BACKGROUND AND DESCRIPTION

Cisco International moved 76% of its orders online, which is equivalent to $28.1 million daily (Cisco Fact Sheet, 2000). It is of course difficult to separate how much of this gain can be attributed directly to *'better ways of doing business'* provided by e-commerce applications versus the gains that have come from the growth of trading partner trust relationships. Figure 1 demonstrates Cisco's history of growth towards e-business. In 1984 Cisco NZ made an initial start and in 1994 a Local Area Network (LAN) and a Wide Area Network (WAN) were introduced. This made them ahead of other international organizations, and by 1998, Cisco was an end-to-end solutions provider. In 1999, Cisco NZ implemented a single network architecture for data, voice and video. One main reason for this success is top management commitment that provided the encouragement and financial resources. Cisco NZ claims that it is important to develop an e-commerce strategy, which complements a corporate strategy. By viewing Internet commerce as a strategic business tool, organizations can support their overall business objectives and profitability goals.

Cisco NZ's e-commerce extranet application called 'Cisco Connection Online' (CCO) was implemented at Cisco International, San Jose, California, USA. Cisco Connection Online provides direct access to manufacturing systems, so that channel partners (trading partners) can track inventories, engineering changes, shipment status and other information in near real time. Cisco Connection Online also provides access to product and marketing information, software downloads and sales tools. The process involves detecting faults and allowing Cisco's trading partners' to download product, equipment and pricing information. By undergoing an automated checking mechanism, 80% of the fault activities were reported on the Web. This online ordering application (CCO) dramatically reduced the costs of sales, distribution, marketing and administration, thereby contributing to savings in administrative costs, telephone calls and delays in responding to queries. The primary business transactions include purchase order of equipment, delivery and product information. Eighty percent of the nontechnical support questions were answered online through CCO convenient, self-service applica-

Figure 1: Cisco's Growth in E-Business (Cisco's Sales Presentation)

Cisco's History

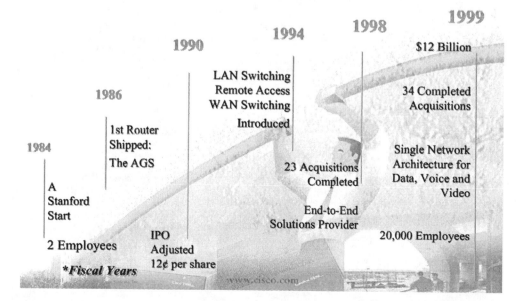

tions. As a result of this, customer satisfaction increased, as they have immediate round-the-clock access to richer, more precise service and support information (Cisco, 2000).

Other secondary elements of CCO included ordering of equipment, delivery and ability to check lead track time. Cisco NZ's trading partners have shown an ability to be competent channel system integrators using Cisco computer and communication equipment thus contributing to competence trading partner trust. Figure 2, below, demonstrates the functions and processes embedded within Cisco Connection Online. The diagram exhibits an online ordering process. Trading partners can check on the pricing and configuration of the order even before placing the actual order. Thus, improved order accuracy from interactive, Web-based applications with built-in rules and access to current pricing, product specifications, selection/configurations and other information ensure the submission of complete and accurate orders. A more efficient and accurate online ordering tool reduces much of the traditional cycle time for handling requisitions and purchase orders, thereby decreasing delivery times to customers and business partners. Cisco NZ claims that 70-80% of Cisco's business involves e-commerce. The annual monetary value from e-commerce

Figure 2: Supporting Tools and Infrastructure of Cisco Connection Online (Cisco Sales Presentation)

Cisco's Full-Service Internet Commerce Implementation

Lead Times

Pricing

Configuration

Order Placement

Order Status

Service Order

Invoice

BEFORE YOU ORDER
- Pricing Tool
- Configuration Tool
- Lead Times Tool

ORDER
- Ordering Tool/IPC (Internetworking Product Center)

AFTER YOU ORDER
- Order Status Tool
- Invoice Tool
- Online Order Extract Tool
- Aged Account Summary Tool

INFRASTRUCTURE ITEMS
- Entitlement/Security
- Secure Transport Architecture
- Globalization
- Data Mirroring
- Data Cleansing
- Partner Initiated Customer Access

SERVICE AND WARRANTY
- Service Contract Tool
- Service Order Submit Tool
- RMA/Service Order Status Tool

- RMA/Service Order Parts Tool
- Service Contract Center Tools
- Returns Tool
- Return Status Tool
- Product Upgrade Tool

OTHER COMMERCE TOOLS
- E-mail Notification Tool
- Billing Address Change Tool

- Core Corba Framework
- Auto Registration
- Customer Service Survey
- Internet Commerce FAQs
- Bug Enhancement Request Tool
- Communications Architecture

transactions is NZ$17-34 billion dollars (US$8-16 billion dollars) contributed from 2.5 million e-commerce transactions per annum. Cisco NZ perceives its organization to engage in long-term business investments with its trading partners (Cisco Fact Sheet, 2000).

Figure 2 shows the infrastructure behind Cisco Connection Online. For example ordering applications is only one part of the CCO. There is a vast portfolio of customer service and technical support applications that include a software download tool kit and various troubleshooting engines that allow customers to enter—in plain-language—problem descriptions into a database.

CCO is Cisco NZ industry-leading online support and information service, available 24 hours a day, seven days a week. CCO provides its trading partners with a wealth of up-to-date information, with hundreds of new documents being added or updated each month. This service is the basis of Cisco's philosophy of moving beyond traditional business barriers that aim to:

- make all of Cisco's information, services and support available to its global customers, partners and employees, on demand;
- deliver faster problem response;
- improve user productivity; and
- significantly lower the cost of doing business.

Thus, implementing CCO can quickly lead to the following compelling benefits:

- increased revenues;
- increased customer and employee satisfaction;
- reduced operating costs; and
- improved productivity of employees, customers and channel partners.

CCO provides Internet-based technical support, by resolving 200,000 telephone calls globally per day without any human intervention. This saves Cisco International $2 million each month. The most significant elements of CCO are the online trading partner support service and networked commerce. The training and support section provides trading partners with online self-help guided assistance. Registered trading partners can log on anytime to access various tools from the company's databases—from intricate details about a particular product or networked environment to fix bugs and update software. A key feature of CCO is its tight and secure integration with Cisco's intranet, which spans 150 locations worldwide. For example in the software library section, customers can find the upgrades and utilities they need, encountering over 16,000 downloads per week. In addition there are innovative tools that help customers to locate the exact information, fixes or troubleshooting tips that they are looking for. Almost half of all trading partners' queries in the U.S., including those from companies like GE and Sprint, are now handled through the Cisco International website which is connected via intranets to Cisco NZ. This helps Cisco NZ to avoid backlogging of telephone-based support calls. According to Cisco NZ, 60 to 70% of all inquiries that come into CCO result in users finding an answer. Cisco NZ's participant indicated that "without CCO, my staff would have to about three times larger to handle the same workload." The Internetworking Products Center (IPC) is only accessible to Cisco's registered direct customers and channel partners, and takes the difficulties out of ordering configurable products by providing an intuitive, easy-to-understand interface. These applications provide order status information, pricing and configuration details. The customers can detail the purchase order numbers, order date, expected shipment date and shipping carrier without calling Cisco's NZ office. They can even link up directly with the Federal Express parcel tracking website from within CCO to find out exactly where their order is; whether it is in the warehouse, on the truck, plane or receiving dock. This type of precise order tracking prevents all kinds of possible billing and shipping problems and provides accurate proof of delivery. It makes communication with Cisco NZ and their trading partners clearer, thereby building trading partner satisfaction and improving trading partner trust relationships.

Cisco NZ is a small-medium sized organization with 20 employees in the Wellington branch. Their reach is international and the product line is data and communication. Cisco International has a sales volume of US$480 billion, with 25,000 employees worldwide, but Cisco NZ, established seven years, has only 20 employees.

Compaq NZ is a large company with 300 employees and their reach is both national and global. They are one of the channel partners (that is the buyer) of computer equipment and data communication parts from Cisco NZ (who is the supplier). Compaq has five branches in New Zealand—one in Wellington, Auckland, Christchurch, Hamilton, and Dunedin. Compaq NZ also obtains directions from its headquarters in the USA. Compaq NZ supplies computer systems, and provides computer services (that is, they are system integrators). Compaq NZ sells computers, undertakes systems integration, application software, hardware, networks, databases and develops database application systems. The network sales specialist indicated that "Compaq NZ's main role is to manufacture computer systems, integration parts and provide computer services." Table 1 summarizes the background information of Cisco NZ and Compaq NZ inter-organizational-dyad. The next section discusses the impact of trading partner trust in business-to-business e-commerce inter-organizational relationships.

Table 1: Background Information of Cisco NZ–Compaq NZ Inter-Organizational-Dyad

Demographic Items	Cisco Limited in NZ	Compaq Limited in NZ
Number of Participants that responded in this study	4	6
Size of Organization	SMEs	Large
Number of Employees	14	300
Main Role of Organization	Supplier and manufacturer-service organization	Buyer
Type of Industry & product line	Computer & Communications	Computer & Communications system integrators
Type of E-Commerce Application	Extranet – application developer (Cisco Connection Online – CCO)	Logs into Cisco's extranet application (CCO)
Number of years using Cisco Connection Online	4 years since late 1996, but had prior trading partner relationships with Compaq NZ	4 years since late 1996, but had prior trading partner relationships with Compaq NZ
Types of business to business transactions	Purchase order of equipment and delivery, information from the web sites, ordering and equipment	Purchase order of equipment and delivery, tracking information from Cisco's wcb sitc.
Number of Trading Partners (using CCO)	20	50 but only Compaq uses CCO

Theories of Trust in Business Relationships

Previous scholars who examined trust in business relationships have identified trust to be a key factor for successful long-term trading partner relationships (Ring & Van de Ven, 1994). For example, trust has been found to increase cooperation, thus leading to communication openness and information sharing (Cummings & Bromiley, 1996; Doney & Cannon, 1997; Morgan & Hunt, 1994; Ring & Van de Ven, 1994; Smith & Barclay, 1997). Furthermore, Granovetter, (1985) suggests that the density and cohesiveness of social networks within relationships influence the evolution of trust. Table 2 demonstrates antecedent trust behaviors and characteristics from previous research that paved the way to the development of three types of trading partner trust.

Competence Trust

Competence trust emphasizes the trust in trading partners' skills, technical knowledge and ability to operate business-to-business e-commerce applications correctly. Trading partners who demonstrate skills and ability in producing high-quality goods and services, such as timely delivery of accurate information to other trading partners, help maintain their supply chains and make strategic decisions that achieve high levels of competence trust (Mayer et al., 1995). Thus, competence trust develops into an economic foundation, and perceived benefits such as savings in cost and time from accurate transfer of e-commerce messages are achieved. Alternatively a lack of competence trust may lead to additional costs, as trading partners need to

Table 2: Different Types of Trading Partner Trust

Source	Competence Trust Economic Foundation	Predictability Trust Familiarity Foundation	Goodwill Trust Empathy Foundation
Gabarro (1987)	Character Role competence	Judgement	Motives/ Intentions
Mayer, Davis & Schoorman (1995)	Ability	Integrity	Benevolence
McAllister (1995)	Cognitive	Cognitive → affective	Affective
Lewicki & Bunker (1996)	Deterrence/ Calculus	Knowledge	Identification
Mishra (1996)	Competence	Reliability	Openness Care Concern

spend time training and educating themselves, in addition to re-sending the same transaction correctly again.

Cisco participants indicated that the competence trust of their trading partners was low. One possible explanation for this is that *"we have some trading partners who can perform competently, and some who cannot, but have to learn the painful way in terms of wasted time, and costs, as they need to re-send the same order twice...Some of our trading partners do not have the underlying fundamental knowledge to place complete and correct orders, thereby lacking the intellectual horsepower required to undertake the ordering process, which is a complex one."* Trading partners will need to become familiar with our products, which demands an ability to configure the orders correctly and completely. Fortunately, Cisco NZ's Internet Business Solutions Group (IBSG) located in Sydney (Australia) provides technical support and handles all queries relating to technical clarifications (Info-Tech, 2000).

Predictability Trust

Predictability trust emphasizes the trust in trading partners' consistent behaviors that provide sufficient knowledge for other trading partners to make predictions and judgments due to past experiences (Lewicki & Bunker, 1996). McAllister (1995), suggests that we choose who to trust and under what circumstances. This choice is cognition based (interpersonal trust), thus investigating past measures of trust, such as reliability and dependability. Perceived benefits include trading partners' satisfaction and information sharing developed from competence trust. Thus, a chain of consistent positive behaviors create a foundation of familiarity, which makes the perception of trading partners reliable, predictable and therefore trustworthy. Alternatively, opportunistic behaviors, such as imbalance of power increase the price of goods or create a demand for high-quality services.

A Cisco participant stated, *"predictability trust in our trading partners to be high, as over time consistent behaviors in their ability to place orders online was observed."* This contributed to two types of loyalty/trust. First, Compaq NZ end customers relied on Compaq as system integrators, and secondly, Compaq NZ became dependent on Cisco NZ's products. By showing consistent behavior in Cisco NZ business interactions (as in providing fast responses to the queries, fixing problems and inquiries on orders and pricing), Cisco NZ was able to develop predictability trust. Trading partners could check on the prices, request for a discount if necessary and electronically obtain an estimated time of arrival before even confirming the order, thereby enabling them to make strategic decisions.

Goodwill Trust

Goodwill trust emphasizes the trust in trading partners' care, concern, honesty and benevolence that allow other trading partners to further invest in their trading partner relationships, thus leading to a foundation of empathy (Mayer et al., 1995). When reliability and dependability expectations are met, trust moves to effective foundations that include emotional bonds, such as care and concern. Goodwill trust is characterized by an increased level of cooperation, open communication, information sharing and commitment, thus leading to increased e-commerce participation. Perceived benefits, such as long-term investments and building the reputation of trading partners, are achieved from goodwill trust. Alternatively, an absence of goodwill trust may lead to termination of trading partner contracts and in some cases a bad reputation among trading partners.

Cooperation determines an organization's willingness to collaborate and coordinate its activities in an effort to help both organizations achieve their objectives, and is defined by the degree to which trading partners cooperate to reach their objectives and make their relationship a success. Cooperation among trading partners reduces conflict, increases communication and enhances trading partner satisfaction (Anderson & Narus, 1990). Cisco NZ's trading partner contracts last between three to five years, and by engaging in long-term trading partner relationships, trading partners are able to increase their volume and dollar value of e-commerce transactions, thereby yielding high profits and achieving satisfaction. Satisfaction in turn increases Compaq NZ's level of commitment, as they were able to realize strategic benefits. Hence, well-planned partnerships with established goals help to build trust which is central to building long-term trading partner relationships, as it reduces the need for extensive control safeguards and paper trials normally absent in e-commerce linkages (Dwyer, Schurr & Oh, 1987; Ganesan, 1994; Morgan & Hunt, 1994).

Compaq NZ was willing to share information regarding the amount of stock they require for an advanced period of time, as they were aware of the estimated arrival dates of the goods they ordered. This enabled Compaq NZ to inform their end customers, thus fulfilling their business promises and building their reputation. Hence, the cyclical process of developing trading partner trust contributed to Cisco NZ's reputation, as Compaq NZ continued to order Cisco products.

Cisco NZ's accounting manager stated that, *"Our trading partners typically tell us that doing business with us is better than doing business with our competitors"* due to the high-quality services that Cisco NZ provides and the reputation that Cisco International holds. For example, the Compaq NZ e-commerce coordinator stated, *"We believe that Cisco NZ staff had the ability to do their job, as they are the 'pros' and they know what they are doing...Their IT support people are excellent, very responsive, timely and professional.*" Figure 3, below, demonstrates Cisco's growth when compared to their competitors. It can be seen that Cisco International has invested in Internet business solutions since 1994, and by 1999 the gap between Cisco International and their competitors was almost $10 billion (Cisco Fact Sheet, 2000). A Compaq NZ participant defined trust as "the information Compaq NZ divulges to their trading partners must be kept confidential, and their trading partners must in turn treat them equally as other business partners." Cisco NZ has a big commitment to make things work. Communicating with e-commerce transactions is not the issue, but what is more important is handling business management issues, which are directly related to the trading partners. For example, giving someone else a bit of price information that will affect the privacy of business information is a concern. Therefore, another Compaq NZ participant indicated that "trust refers to both in meeting the operational technical needs, and more importantly the needs of the trading parties themselves."

Cisco International saves up to US$800 million in costs per year from online ordering, as there is no need to re-key the same order information that was entered by their trading partners (Cisco, 2000). Training is given on how to

Figure 3: Cisco's Growth in Line with its Competitors
(Cisco Sales Presentation Slides)

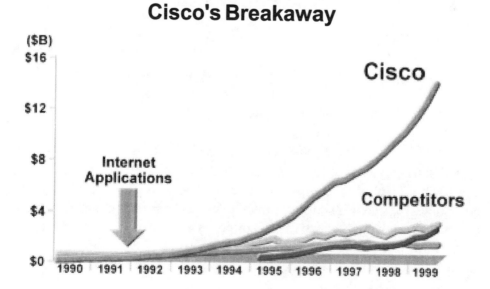

use the extranet application, thus leading to savings in time and costs from reduced error rates, and improved accuracy of information exchanged. For example, the error rate for Cisco NZ of 80% in 1997 has reduced to less than 15% in 2000. The provision of real-time, online tracking information to their trading partners via Cisco Connection Online applications has contributed to additional savings in time and costs. These benefits further contributed to an increase in e-commerce participation. For example, competence trust derived from efficiency benefits of Cisco Connection Online concentrated on reducing transaction costs, from speed and automation of e-commerce applications. Economic benefits over a period of time led to positive consistent behaviors from Compaq NZ thus leading to personal benefits. Personal benefits include improved customer service, product quality, satisfaction, improved productivity and profitability, as costs was no longer spent in fixing errors, thereby leading to competitive advantage and strategic benefits.

CURRENT CHALLENGES/PROBLEMS FACING THE ORGANIZATION

The findings of the case studies contributed to increased awareness of the importance of trust, thereby helping organizations to design more effective strategies, trading partner agreements and partnering/relationship charters. Cisco NZ claims that it is important to build and maintain positive trading partner trust relationships, in order to increase e-commerce participation. Hence, in order to remain competitive, businesses must have a strategy for sales and support over the Internet. In today's hyper-competitive global marketplace, the pressure is increasing from customers and shareholders to provide easy-to-use online applications, as a better way to conduct business. For most organizations, the biggest challenge is not if or when to consider an Internet commerce solution, but rather how to select the best Internet commerce strategies and tactics to develop and sustain competitive advantage. Cisco NZ's success has contributed to the following e-commerce strategies:

Well-Planned Partnerships

Cisco NZ's trading partners were chosen on the basis of their reputation, and by replicability as channel distributors. By contrast to many technology companies, Cisco NZs does not take a rigid approach that favors one technology over the alternatives and imposes it on their trading partners as the only answer. Cisco NZ's philosophy is to listen to their trading partners' requests, monitor all technological alternatives and provide trading partners with a range of options from which to choose. For Cisco's e-commerce strategies to work, it was important for Cisco NZ to have good relationships with its suppliers, factories and the companies that deliver Cisco's products, as well as their customers.

In addition, Cisco NZ embraces the Global Networked Business model, which aims to implement innovative tools, systems and share information with diverse company stakeholders. Thus, Cisco NZ's shared commitment with their strategic alliances to deliver solutions and services was designed to help deliver a customer-centric, total solution approach to solve problems, exploit business opportunities and create sustainable competitive advantage for their trading partners. This shared commitment helped deliver solutions, products and services, together with applications on systems integration and best practices that made Cisco's trading partners successful as globally networked organizations in the new economy.

Online Support

Cisco NZ has investments in regional technical assistance centers based in San Jose, Sydney, Belgium and North Carolina. Technical support for complex Cisco NZ's products is still delivered by humans via telephone and computer. Cisco NZ's Internet Business Solutions group based in Sydney (Australia) collaborates with their trading partners in order to design and build applications and networks that optimize the e-business software.

Trading Partner Satisfaction

Cisco NZ achieves 100% order accuracy by doing everything online. Errors were inevitable when Cisco NZ had human involvement in the ordering process. Trading partners (Compaq NZ) would often submit orders for products that might not work, thus causing delays and eroding customer satisfaction. Now they can go to the website and configure the products online and be aware of what they are ordering. If not, the system will recommend the right configuration. Asking trading partners to do their own ordering might appear to be an erosion of service, but the opposite is true. Trading partners have full control from the moment they place an order with Cisco NZ. They can even trace order shipments online, through Web links to Cisco's delivery partners such as Federal Express.

Best Business Practices

Cisco NZ's best business practices included strategic marketing to gain a complete analysis and understanding their trading partners' behavior. For example, their experience with Compaq NZ initiated with a number of negotiations which led to capacity planning that applied *"just enough infrastructure"* and policies to optimize capital expenditures. The Cisco Connection Online system consists of a highly reliable, scalable and distributed architecture, which delivers Internet audio, voice and text applications to any telephone. The Web included user-friendly browsers and open infrastructures that enabled content and application providers to develop e-commerce. Specifically, the Web gives Cisco NZ a vehicle through which trading partners can find out about products and buy, along with an automated support system that can reach a larger audience. A certification program further assures functionality of Cisco NZ technologies and full interoperability of the device in heterogeneous networks; Cisco's network infrastructure was also implemented in order to ensure assurance. Furthermore, fraud management using real-time detailed user profiling to spot unauthorized users, non-billed use and excessive data-storage problems. Thus, effective trust and security-based mechanisms were imposed, in order to protect the confidentiality, integrity, authenticity, non-repudiation and availability of business-to-business e-commerce transactions.

Universal Standards

Cisco NZ develops its products and solutions around widely accepted international industry standards. Cisco NZ abides by the standards and procedures from its headquarters. In some instances, technologies developed by Cisco International have become industry standards themselves. Cisco NZ describes this change as the global networked business model and refers to a global network business as an enterprise, of any size, that strategically uses information and communications to build a network of strong interactive relationships with all its key constituencies. In addition to its worldwide leadership in networking for the Internet, Cisco is a global leader and industry benchmark for Internet commerce. Hence, Cisco NZ is on the cutting edge of using networks to leverage its business relationships.

Increased Employee/Trading Partner Productivity and Satisfaction

Cisco NZ's online systems were designed to be user friendly, easy to use and interactive. Their intranet Web-based e-commerce applications enabled more efficient processes for conducting transactions online with their trading partners. By automating many of the routine and administrative sales, order and customer service functions, employees are able to improve their productivity and focus on more challenging and rewarding interactions. Online tools allow trading partners to access support and order information online, rather than through a Cisco sales representative.

Cisco NZ's experience in e-commerce has set the standard for e-business transformation, creating Internet solution leading practices in the transformation of core processes, thus building their reputation. Cisco NZ's established technology architecture and leadership utilizes intelligent network services to offer a complete end-to-end solution. Cisco has partnered with the best of breed application providers and system integrators (Compaq NZ), who have proven experience building e-business applications and solutions. This e-commerce pioneer can show other companies how to plan, execute, manage and improve their infrastructures for Web-based sales.

The online ordering applications dramatically reduced the cost of sales, distribution, marketing and administration. Cisco Connection Online gave Cisco NZ higher gross margins and happier customers. Hence, it was confirmed by Cisco NZ participant that the success of e-commerce depends on well-planned partnerships, and the need for effective internetworking. The teaching case developed for this study introduces some questions relating to inter-organizational relationships in business-to-business e-commerce participation.

REFERENCES

Anderson, J. C. and Narus, J. A. (1990). A model of distributor firm and manufacturer firm working partnerships. *Journal of Marketing,* 54(January), 42-58.

Cisco. (2000). *The Global Networked Business: A Model for Success.* Retrieved on the World Wide Web: http://www.cisco.com/warp/public/756/gnb/gnb_wp.htm.

Cisco Fact Sheet. (2000). *Cisco Systems is the Worldwide Leader in Networking for the Internet.* Retrieved on the World Wide Web: http://www.cisco.com/warp/public/750/corpfact.html.

Cisco Newsroom. (2001). Cisco Systems Incorporation Announces Cost Cutting Measures. *Measures Address Changes in the Global Economy.*

Cummings, L. L. P. and Bromiley. (1996). The organizational trust inventory (OTI): Development and validation. In Kramer, R. M. and Tyler, T. R. (Eds.), *Trust in Organizations: Frontiers of Theory and Research,* Thousand Oaks, CA: Sage Publications, 302-220.

Doney, P. M. and Cannon, J. P. (1997). An examination of the nature of trust in buyer-seller relationships. *Journal of Marketing,* April, 35-51.

Dwyer, R. F., Schurr, P. H. and Oh, S. (1987). Developing buyer-seller relationships. *Journal of Marketing,* 51, 11-27.

Gabarro, J. (1987). *The Dynamics of Taking Charge.* Boston, MA: Harvard Business School Press.

Ganesan, S. (1994). Determinants of long-term orientation in buyer-seller relationships. *Journal of Marketing*, 58(April), 1-19.

Granovetter, M. (1985). Economic action and social structure: The problem of embeddedness. *American Journal of Sociology*, 91(3).

Hart, P. and Saunders, C. (1997). Power and trust: Critical factors in the adoption and use of electronic data interchange. *Organization Science*, 8(1), 23-42.

Info-Tech Weekly. (2000). Cisco chalks up $64m in sales a day. February 14.

Info-Tech Weekly. (2000). E-business stories. August 23.

Keen, P.G.W. (1999). *Electronic Commerce: How Fast, How Soon?* Retrieved on the World Wide Web: http://strategis.ic.gc.ca/SSG/mi06348e.html.

Lewicki, R. J. and Bunker, B. B. (1996). Developing and maintaining trust in work relationships. In Kramer, R. M. and Tyler, T. R. (Eds.), *Trust in Organizations: Frontiers of Theory and Research*, Thousand Oaks, CA: Sage Publications, 114-139.

Mayer, R. C., Davis, J. H. and Schoorman, F. D. (1995). An integrative model of organizational trust. *Academy of Management Review*, 20(3), 709-734.

McAllister, D. J. (1995). Affect- and cognition-based trust as foundations for interpersonal cooperation in organizations. *Academy of Management Journal*, 38(1), 24-59.

Mishra, A. K. (1996). Organizational responses to crisis: The centrality of trust. In Kramer, R. M. and Tyler, T. R. (Eds.), *Trust in Organizations: Frontiers of Theory and Research*. Thousand Oaks, CA: Sage Publication, 261-287.

Morgan, R. M. and Hunt, S. D. (1994). The commitment-trust theory of relationship marketing. *Journal of Marketing*, (58), 20-38.

Ring, P. S. and Van de Ven, A. H. (1994). Developing processes of cooperative inter-organizational relationships. *Academy of Management Review*, 19, 90-118.

Smith, J. B. and Barclay, D. W. (1997). The effects of organizational differences and trust on the effectiveness of selling partner relationships. *Journal of Marketing*, 51, 3-21.

BIOGRAPHICAL SKETCH

Pauline Ratnasingam is an Assistant Professor in the School of Business Administration, The University of Vermont, Burlington, Vermont. Before that she was a Lecturer at the Victoria University of Wellington, New Zealand. Her Ph.D. dissertation examined the importance of inter-organizational trust in electronic commerce participation (the extent of e-commerce adoption and integration). Her research interests include business risk management, electronic data interchange, electronic commerce, organizational behavior, inter-organizational relationships and trust. She has published several articles related to this area in conferences and refereed journals. She is an associate member of Association of Information Systems (AIS).

Geographical Information System (GIS) Implementation Success: Some Lessons from the British Food Retailers

Syed Nasirin
Universiti Sains Malaysia

EXECUTIVE SUMMARY

Geographical Information Systems (GISs) are becoming more prevalent for retailers in their use for both day-to-day and strategic long-term decision-making. Given the array of internal and external databases they use, as well as the amount of organizational development, systems implementation is a most opposite picture of how GIS support retailing decision-making. This chapter presents the results of in-depth case studies, reflecting upon the GIS implementation experiences of a key UK food retailer, Highway Stores PLC. The company is one of the strongest contenders in UK food retailing (fourth in rank). More sites throughout the country are being explored and considered for development, and the implementation of GIS has supported Highway in determining locations where new stores can be built-up that fascinate new customers and ensure that existing customers are retained.

BACKGROUND

Highway, the fourth largest food retailer in the UK, is a subsidiary that belongs to the Ali group of companies. As part of the Ali group since 1987, Highway has become the group key retailing face (accounting for more than 80% of the group retail sales). Throughout the country, Highway had a total of 490 stores comprising 400 Highway stores and 90 Presto stores. In 1996, Highway accounted for about 92.5% of Ali's operating profit and turnover. As a result, Highway's Board of Directors decided to change its name from Ali to Highway Plc. Highway has had a successful period in terms of sales growth. In the 1991-1996 period, it has achieved the third highest growth rate, beating J. Sainsbury. During this period, Highway's sales performance has been inspiring (whose sales area in 1991 was less than three-quarters of that Asda). It ranked second after Asda in terms of sales density growth. A key factor behind the rising sales was, however, increase in sales footage.

Competition, Market Share

Key food retailers make up the bulk of Highway's competitors. Highway battled with other large food retailers while facing a high degree of competition as one of the leaders in the industry. Table 1 shows Highway's performance in terms of market share and operating profit among other key food retailers. Both sets of data show the same trend, notably the ever-growing concentration of power among the key grocers (focusing its efforts on fulfilling market needs in evolving market segments). Moreover, Table 2 shows the performance of the UK's top four grocers.

Although Highway's average store size has grown, it remains smaller than that of its key competitors, with the exception of Waitrose.

Table 1: Highway's Market Share and Operating Profit Performance Amongst Other Key Food Retailers (1997)

	1997	
	Market Share (%)	Operating Profit (£M)
Tesco	23.6	760
J. Sainsbury	19.6	661
Asda	13.5	365
Highway	*10.8*	*410*
Kwik Save	5.8	74
Somerfield	4.5	115
Wm Morrison	4.0	134
Iceland	3.0	65
Waitrose	1.6	74

Table 2: The UK Top Four Food Retailers' Performance (1997)

	Turnover (£M)	Operating profits (£M)	Operating profit margins (%)	Sales per square foot per year (£)	Operating profit per square foot per year (£)
Tesco	13,118	760	5.8	935	54
J. Sainsbury	10,852	661	6.4	1,045	67
Asda	6,883	365	5.3	713	42
Highway	*6,590*	*410*	*5.9*	*655*	*48*

SETTING THE STAGE

In the UK, for the last 25 years GISs have rapidly developed as Decision Support Systems (DSS), notably employed by retailers are becoming increasingly crucial to support both operational day-to-day and strategic long-term decisions. A GIS is defined as:

"An emerging science of spatial information, it deals with how to collect, compile, store, analyze and display spatial data within a digital environment, raising explicit questions that have previously always been implicit within spatial analysis, such as the measurement of accuracy of spatial data." (NCGIA[1], 1989)

In retailing, GIS is also known as "geodemographics." It is derived from the combination of both geographic and demographic terms. The system was initially employed to support site selection decisions, but have developed to support an array of marketing mix decisions. As GIS "re-engineers" the traditional working approaches and involves continuous commitment from all the parties (senior managers, system developers and users) in the organization, the system fundamentally changes the existing organizational working approach towards site selection and other marketing mix decisions.

Given the scope and flexibility of a GIS, more and more retailers of various types are employing the system, examples being Boots the Chemists, Marks and Spencer and W. H. Smith. The need for retailers to analyze their market has grown as the competition faced by the retailers becomes more intense, for example, the opening-up of new markets, particularly in the European Community. This situation further justifies why retailers are changing their focus towards GIS technology.

Its relative advantage lies in its ability to locate the customers through its discriminatory power, i.e., the system is capable of merging various retailers' internal and external databases. This integration has allowed site researchers to make full use of the existing datasets. The availability of GIS databases at a national level—for example, in the UK, the Target Group Index (TGI) and National Readership Survey (NRS)—is also one of the continuing factors in changing retailers' perspectives towards this evolving system. These databases are also becoming more portable at a reasonable cost (from mainframe to CD-ROM). The data will be more costly if it is self-acquired by the retailer.

In short, GIS has been employed by retailers in all sorts of marketing mix decisions, e.g., direct mailing, such as door-to-door distribution of leaflets through postcodes clusterings, selecting newspaper readers and television audiences, merchandise management and sales forecasting. Such systems are established as giving competitive advantage, enhancing organizational planning and decision-making in a wide array of functions.

CASE DESCRIPTION

GIS Implementation Process at Highway

Operating in a mature market where there were many competitors offered a challenge in terms of analyzing which stores should close and where new ones should be opened. The Stores Information Department (SID) key function was to advise Highway on where the organization should invest in new sites. An average of 25 key decisions were made every year on site selection for new stores. In addition, SID also provided advice on the performance of existing stores. As one of the top food retailers, SID was responsible for monitoring the 490 Highway stores. There were about 16 site researchers working with the department. Figure 1 illustrates the SID organizational structure.

Figure 1: Stores Information Department (SID)'s Organisational Structure

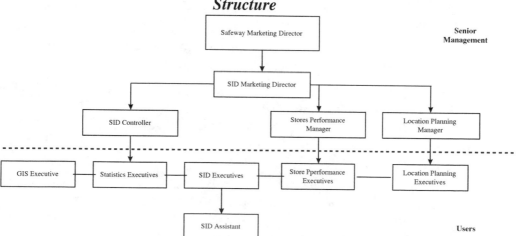

Much work has been done on streamlining the stores by eliminating the poor performers. Highway reduced its number of stores while raising its total selling space (see Table 2). Furthermore, during the 1995-96 period, 20 Highway stores were closed. The development trend was to have a relatively small store size. Although the average size of the stores was smaller than most of its competitors, Highway managed to cram in a wide range of services, including coffee shops, dry cleaners, petrol stations, pharmacies and even post offices in many of their larger stores. These were seen as significant complimentary services to their main store offerings.

Prior to the implementation of the system, much time was spent by SID managers in manually evaluating Highway's potential sites and their catchments through the overlaying of the available spatial information, e.g., Ordinance Survey maps and census data. This traditional approach to site selection decisions inherited three significant problems:

- The external (for instance, National Shoppers Survey) and internal (for instance, EPOS) data sets were somewhat in disarray as the amount of both data increased, so did the difficulties of storing the data.
- The outputs of traditional approach were in the form of non-graphical data. To graphically represent and analyze potential sites, SID had to manually place different color pins on paper-based maps.
- Senior management pressure to hold its current position in the market (one of the company aims) forced SID to shorten its site selection process, as the traditional approach was somewhat deliberate.

At Highway, two basic reasons for investing in GIS were;

- That GIS would lead to a productivity increase by expediting the site selection process, e.g., more sites could be analyzed and selected for expansion.
- That GIS would save money by automating the collection and storage of surveyed data for use in site selection decisions.

The first system used by SID was a PC-based GIS (stand-alone), which ran on an MS-DOS platform. The system was primarily undertaken for operations support to the site selection decisions. It represented significant operational tasks vital in the day-to-day running of the department. As the amount of data increased, the system was incapable of managing effective existing databases in which greater volumes of data were flowing into the department. The system was also incapable of providing customized SID internal site selection needs, i.e., the requirement for more rigorous and sophisticated analysis (an increasing significant feature as GIS applications had developed). In addition, the SID Director was increasingly frustrated at his department's inability to accurately select new sites for Highway's stores because with the traditional approach of analyzing potential sites, there was always a backlog of surveys to analyze. As a result, he decided to go for a thorough network-based GIS implementation.

An investigation into the suitability of GIS was started by an enthusiastic champion. A series of GIS implementation discussions were conducted with GIS specialists in non-competing organizations outside Highway, examining how they implemented GIS in their organizations, i.e., discussions on the issues faced in designing the databases. Further, by engaging users with the implementation project, SID managers had the opportunity to reinforce the sense of users' commitment and ownership to the GIS through participating in the project's conception. Through presentations and discussions, users at all levels developed a reasonable understanding of what was being built and what was going to be built. Highway GIS had been developed with emphasis upon human-computer-based interfaces that were easily utilized by site researchers, which required minimal support from the Information Systems Department (ISD).

It was not possible to purchase an off-the-shelf GIS solution. This was because off-the-shelf GISs were unable to be customized. The system chosen was an object-oriented (OO) network-based GIS by Laser-Scan (known as Market Analysis) which ran on the organization's IBM workstations. It was chosen to be the backbone of the Highway GIS. The applications were developed by both Laser-Scan and SID specifically in location planning and other marketing activities, e.g., promotions and product development. The flexibility of this system was described by one of the SID managers as:

"Laser-Scan's solution gives us the flexibility to analyze spatial data against a geographical backdrop and thereby make the optimum decisions."

A conceptual framework of the entire GIS implementation process was established after a series of discussions among SID managers, and through the wide involvement of managers and site researchers in various job roles, a team was able to develop the system. The purpose of this framework was to create a shared understanding of the general GIS once it was operational. Its ultimate objective was to implement a system that met all critical departmental business needs. The database, for instance, servers and workstations were accessible throughout the various departments and were associated with remote posts, through distributed terminals. GIS technology was used throughout this period and was recognized as a decision support tool for SID.

At the time of the study, Highway was using Laser-Scan's Strategist to great effect in assisting their location planning activities. Site selection decisions could be achieved in a much shorter period (it had facilitated strategic decision-making and helped to ensure the integrity of the organization, particularly as competitive pressures increase). As the SID Director described, "We see this type of technology as key in maintaining our competitive edge." The introduction of GIS was seen as offering the potential of spreading the applications around other departments within Highway, as it was significant to integrate the GIS into the entire corporate information systems. The interest in the technology opened a good cooperation between SID and other departments. Its openness to exchange data yielded an infrequent but useful communication of mutual benefits. Figure 2 summarizes the SID's GIS implementation process framework.

Further, system developers paid careful attention to users in developing the applications, as they believed that applications, e.g., user interfaces, should accommodate user needs and not the other way around. They had taken a problem-based approach[1] in designing those applications. Application prototypes were developed for the application and database designs to evaluate the system prior to its actual implementation (different applications necessitated a careful prototype design to meet the needs of a particular situation). The experience gained enabled progress to be made towards the

Figure 2: SID's GIS Implementation Process Framework

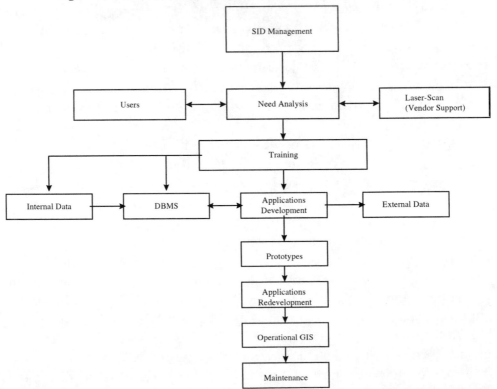

implementation process. These processes demonstrated a real working GIS to the users. Users gained more confidence with the system in which they were able to perform routine site selection decisions.

GIS implementation proceeded slowly with commitment and support from SID management due to the belief that lies in the ability of the technology. Management support was described by one of the users as:

"Initially it was not brilliant, it was not brilliant. It tended to be a real mix of people. Some users are really forward; some executives are really forward but not everybody. So it's a real mismatch. We did not have a broad level of support."

She further added:

"I am very lucky because my boss [SID Director] is the person that has fought for the implementation of the GIS. So he is very interested. It is good to actually have a manager who is really into the GIS."

The SID Director played his role by supporting the project. He made the ultimate decision to purchase a Laser-Scan GIS and led the purchasing negotiations. The purchase was made with the departmental funds with the permission from the Highway Board of Directors. He committed himself entirely to the idea of implementing the system within Highway (he was seen as the key advocate of the system and led the Laser-Scan network-based GIS purchase). Besides him, there were also a few managers and users who also acted as champions. It was widely accepted that without champions, Highway could not successfully implement a GIS. They also spread the news about GIS to potential users in other departments. They tirelessly pursue the goal of GIS implementation and its benefits by "hardly" selling the idea to management and co-users. They also communicated the needs of GIS implementation activities up front and make potential users comfortable and productive when they understood what was expected of them more.

Building a GIS was a matter of constructing graphic and non-graphic databases, developing its processing capabilities, installing the appropriate hardware and software and implementing the procedural changes needed to operate and use the system successfully. These were the essential tasks to be accomplished but could not be started until all participating users knew what they expected the system to do for them. Addressing the users' needs assisted in GIS implementation by encouraging users to participate in this process allowed refinement of Highway's GIS implementation as well as exposing strengths and weaknesses of the system. User issues raised under the prototyping process brought up to the meeting for further discussion, i.e., should a GIS be installed as a common resource to meet all departmental requirements or should it be implemented to provide task-specific applications? It was the responsibility of SID managers that users were identified as being involved in the process.

The SID Director also ensured that anyone whose work was affected by GIS was properly trained. Training strategies were determined by SID managers. There were two modes of training offered by Laser-Scan as the key vendor of Highway GIS:

- *Scheduled Courses*—Laser-Scan ran quarterly scheduled training courses for managers, and users. Individual training sessions range from one to seven days' duration and were held at Laser-Scan's premises on the outskirts of Cambridge.
- *Nonscheduled Courses*—Besides the planned courses, there were also nonscheduled courses. These training sessions were usually conducted at Highway premises.

Users developed considerable expertise and acquiring an excellent reputation for reliability in running the system. A training need analysis was carried out by SID managers to survey users' needs and to ascertain what was the needed training and support. A training assessment was conducted based on the users' tasks (as implementation progressed, there was an attempt by the managers to avoid the "blanket approach" in training users). The SID Director expected that the effort could be repaid by ensuring users had the right training they needed. Training manuals were developed as the implementation process went on by both SID and ISD. The outcomes were described by one of the users as:

"They will have the chance to make mistakes, when that [GIS] is installed on the PC, they are not scared of everything, they are quite familiar."

Database management system (DBMS) was an appropriate tool for effective GIS management. Without it, fast access to internal and external data out of the large amount of operational data collection was difficult to achieve. Besides relying on external data vendors, internal data were converted internally by SID with some help from ISD. Further maintenance, e.g., updates of the databases, was handled internally by SID on day-to-day basis. Data were gathered on a regular basis and fed into the system. The data were available to be distributed so that each site researcher had fast access to it. The loyalty scheme was an attempt to secure a greater proportion of each customer's total spending. The company claimed that its loyalty card has brought them real benefits. By March 1996, 4 million customers had signed up for the Highway's card. Highway believed that much of its 5% profits and sales increase (1995-96) were due to the card. A significant aspect of this scheme was the degree to which the information in the spatial database was maintained. Through their GIS, Highway launched a promotion aimed at encouraging regular card use and recruiting new cardholders to the scheme. Highway planned to use its card database to full advantage, e.g., a link has been established with Lloyds Bank to market insurance products directly to cardholders.

PRESENT CHALLENGES FACING THE ORGANIZATION

Database Management

One of the data issues that SID faced was the problem of the system's base maps, where the accuracy was a critical factor. Perhaps this issue was the most critical issue in the implementation of the GIS project. As the SID Director described:

"The only and the most critical problem faced was the conversion of the maps, the base-maps of the system."

Most of the internally generated data were stored in mainframes, resulting in minimal problems of data conversion. The external data were bought from data vendors and were also stored in mainframes as common GIS data needed by most of users were stored in a central position while individual or other data were kept by users themselves. There was a continuing attempt to keep both internal and external data together in one place. Further, in transferring the data throughout the department, SID used Local Area Network (LAN). The critical issue faced at this point was the "downloading" of huge bytes of map files which were required in almost all the applications.

Resistance to Change

There was also user resistance, as users were somewhat "painful" in using the technology. However, as the implementation activities proceeded, resistance was slowly being overcome by users themselves, through the support provided by their senior managers. In encountering these resistances, the SID Director described his approach as,:

"There has been resistance and it's a bit like sort of rugby playing. All you can do is just bend down your head and keep going."

The design of a GIS can only be as good as the analysis of the need for that system. Prior to the purchase of the system, a survey on user needs was conducted by the SID Director. User need analyses were established within the context in which the system was to be used. Once the "understanding" of users was in hand, the SID Director moved forward by searching for a GIS, which was available in the market that could fulfill these specified needs.

Senior Management Support

As a result of a series of successful presentations to senior managers and users by the champions, GIS was well received and was consequently given a main concern for implementation. Continuous management communications, e.g., electronic-mail announcements and regular departmental meetings, were perceived as essential in smoothing the implementation process as well as clearing the "doubtful thoughts" possessed by senior management and users. As one of the users described:

> "I think management has a very big role to play in helping and guiding you and we know it will take longer to use initially because you are not used to using this. Its quite a radical change in how you do work."

Most users at Highway were "sold" by their superiors on the features of the system; in performing their tasks, due to the small number of staff members within SID, face-to-face communication was used as a major mode of discussion. Various types of support could be seen, e.g., GIS circulars, magazines and manuals were made available by senior managers to help users further understand the technology. Meanwhile, senior managers were also aware of the increasing intricacies in managing the system, e.g., resulting from the increasing amount of data. They believed that the team working spirit was high within their department, which in turn has smoothed the implementation process. There were also a few expert users who were "wandering" around the department to help other users with their queries about the system. In addition, after exposing the system to other influential senior managers (perceived to be possible champions), more discussions were held by the SID Director to promote GIS further.

ENDNOTES

1 Problem-based approach focuses on defining the problem so thoroughly that the appropriate solutions are almost obvious. Central to this approach is the development of a list of performance criteria that defines how the final application should perform.

FURTHER READING

The Analysis, Design and Implementation of Information Systems, (4th Ed.). (1992). New York: McGraw-Hill.

Azad, B. (1992). Case study research methods for geographical information systems. *URISA Journal*, 4(1), 32-44.

Campbell, H. J. (1990). The organisational implications of geographic information systems for British local governments. Paper presented at the *European Geographic Information Systems Conference*, April, 10-13. Amsterdam, The Netherlands.

Campbell, H. J. (1991). *Impact of Geographic Information Systems on Local Government (TRP101)*. Department of Town and Regional Planning, University of Sheffield.

Huxhold, W. E. and Levinsohn, A. G. (1995). *Managing Geographic Information Projects*. New York: Oxford University Press.

Kivijarvi, H. and Zmud, R. W. (1993). DSS implementation activities, problem domain characteristics and DSS success. *European Journal of Information Systems*, 2(3), 159-168.

Kraemer, K. L., King, J. L., Dunkle, D. E. and Lane, J. P. (1989). *Managing Information Systems: Change and Control in Organisational Computing*. San Francisco: Jossey-Bass Publishers.

Lucas, H. C., Jr. (1981). *Implementation: The Key to Successful Information Systems*. New York: Columbia University Press.

Mennecke, B. E., Crossland, M. D. and Killingsworth, B. L. (2000). Is a map more than a picture? The role of SDSS technology, subject characteristics, and problem complexity on map reading and problem solving. *MIS Quarterly*, 24(4), 601-629.

NCGIA. (1989). The research plan of the national center for geographic information and analysis. *International Journal of Geographical Information Systems*, 3, 117-136.

Onsrud, H. J. and Pinto, J. K. (1991). Diffusion of geographic information innovations. *International Journal of Geographic Information Systems*, 5(4), 447-467.

Prerau, D. S. (1990). *Developing and Managing Expert Systems*. Massachusetts: Addison-Wesley.

BIOGRAPHICAL SKETCH

Syed Nasirin is a Lecturer in Retailing in the School of Management, Universiti Sains Malaysia, Penang. He has a Ph.D. from the University of Bath and an MBA from Ohio University. He has also served the Victoria University of Wellington as a Visiting Lecturer in Information Systems. His research focuses on IS/GIS implementation in retailing in the United Kingdom, Thailand, Malaysia, New Zealand and Australia. Nasirin has consulted with many leading multinationals operating in Malaysia including Carrefour. He is now the Chief Editor for the Asian Academy of Management Journal, *a double-blind peer-reviewed journal published by the Asian Academy of Management.*

Implementing and Managing a Large-Scale E-Service: A Case on the Mandatory Provident Fund Scheme in Hong Kong

Theodore H. K. Clark, Karl Reiner Lang and Will Wai-Kit Ma
Hong Kong University of Science & Technology, Hong Kong

EXECUTIVE SUMMARY

This case concerns a recently launched retirement protection scheme, the Mandatory Provident Fund (MPF), in Hong Kong. Service providers, employers, employees and the government are the four main parties involved in the MPF. The service has been implemented in two versions, i.e., a bricks model and a clicks model. The former is based on conventional paper-based transactions and face-to-face meetings. The focus of this case, however, is on the latter, which introduces MPF as a service in an e-environment that connects all parties electronically and conducts all transactions via the Internet or other computer networks. The case discusses the MPF e-business model, and its implementation. We analyze the differences between the old and the new model and highlight the chief characteristics and benefits of the e-business model as they arise from the emerging digital economy. We also discuss some major problems, from both managerial and technical perspectives, that have occurred during the phases of implementing and launching the new service.

BACKGROUND

The case discusses the traditional and the new e-business model in providing retirement management services in Hong Kong, and examines how a Mandatory Provident Fund (MPF) service provider can benefit and gain competitive advantage through the application of current e-commerce technologies. The case provides comprehensive background information on the MPF scheme and the MPF market, and presents the newly developed MPF e-business model and its implementation as an electronic service (e-MPF).

The main purpose of this case is to develop and analyze the main components of this new e-service model, and to understand both its benefits and limitations in today's nascent e-environments (Westland & Clark, 1999). We present the issues and challenges that emerge from introducing and managing a new, large-scale service that involves several parties, each with different objectives and agendas. The government acts as the regulating body formulating the rules and overseeing the transactions between MPF service providers, employers and employees. The government's primary goal in introducing the MPF scheme is to provide basic retirement protection for Hong Kong's workforce in an effort to catch up with other developed countries that provide basic social welfare for its citizens. The service providers, on the other hand, see the emerging MPF market as a huge business opportunity. Employers and employees are forced by law to participate in the new scheme.

MPF and the New Economy

The Mandatory Provident Fund is an interesting business example in the emerging digital economy, and one of the first large-scale e-service projects in the public sector in Asia. The e-MPF model incorporates some fundamental e-business concepts as follows:

Electronic Money
There is no real physical activity or presence required in any of the transactions. Fund transfers, monetary transactions, contractual and regulatory settlements, and information exchanges are all done online.

Digital Distribution
In the traditional MPF model, distribution of paper documents and physical, contact-based customer service accounted for more than 30% of total cost. Moving to digital distribution and document management is expected to significantly reduce distribution cost.

Knowledge Management

In order to retain the best customers, MPF providers should supply their customers the most appropriate product and service information. Informed customers often prefer to make their own decisions in designing their MPF investment plan. The e-MPF business model suggests increasingly active and dynamic investment behavior.

Customer Relations

Being able to understand and anticipate the needs of the customer will be imperative in the new competitive MPF market. All leading MPF service providers will emphasize closer customer relationships as one of their priorities, and will employ modern IT systems to support customer relationship management.

MPF in an Electronic Marketplace

A truly electronic market is emerging for MPF services (Bakos, 1998). The entire value chain, including the post-sales investment asset management, can be conducted in digital space. All market participants, service providers, clients and the government share the same Internet-based business platform. However, the MPF service providers also offer a traditional, paper-based version of their service product. At least in the initial years, it is expected that most MPF clients will prefer face-to-face transactions and paper documents over electronic transactions. Especially low-wage workers and clients with low education are reluctant in using the electronic MPF (or e-MPF) services. But given that Hong Kong has one of the highest Internet penetration and growth rates in Asia and one of the best Internet infrastructures in place, it can be expected that more and more MPF clients will migrate to the e-MPF service platform over the years. Hence, all MPF service providers place strategic importance on e-MPF. Our discussion in this case emphasizes the electronic version of the MPF services.

Hong Kong is widely considered as a highly developed, ultra modern city-state. Its welfare system, however, is lacking. It is the family, not the government, that traditionally provides retirement support in Chinese societies. People with employer-provided retirement coverage accounted only for a minority until the launch of MPF. Retirement schemes were mostly arranged through either insurance or asset management companies. However, in an effort to bring Hong Kong up to par with other developed countries, the government has recently introduced the MPF Scheme, which would provide a safety net to much of its population. The particular scheme developed in Hong Kong is based on a similar scheme in Singapore. However, while the

Singaporean model is strongly government-based, Hong Kong has opted for a more market-based solution. Specific details on the HK system can be found on the Mandatory Provident Fund Authority's official website (MPFA, 2001).

Hong Kong has a rapidly aging population. People aged 65 and above presently account for about 10% of the population. This proportion is estimated to increase to 13% by 2016, and to 20% by 2036. Only about one-third of the workforce has had some form of retirement protection. After decades of debates on the provision of retirement protection, Hong Kong took a major step forward in August 1995 when the Mandatory Provident Fund Schemes Ordinance was enacted to provide a formal system of retirement protection. The Mandatory Provident Fund Schemes Regulation was passed on April 1, 1998. The Mandatory Provident Fund Schemes Authority set December 1, 2000, as the enrollment deadline and for starting the collection of the contributions. From the period February to October 2000, HK's 251,000 employers had to complete the selection of an MPF service provider and arrange for employees to join the MPF scheme. The government set December 1, 2000, as the deadline for employers and employees to enroll in one of the approved MPF schemes.

The MPF System

The MPF System is an employment-related contributory system, under which members of the workforce aged between 18 and 65 are required to participate. There are three main types of MPF schemes:
1. master trust scheme,
2. employer-sponsored scheme and
3. industry scheme.

Figure 1: Population Statistics

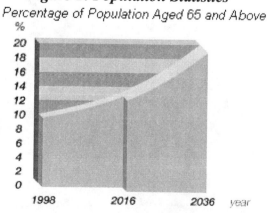

Percentage of Population Aged 65 and Above

A master trust scheme is a scheme open to membership to the relevant employees of more than one employer, self-employed persons and persons with accrued benefits transferred from other schemes. This type of scheme is especially suitable for small and medium-sized companies.

An employer-sponsored scheme is a scheme that is only open to the eligible employees of a single employer and its associated companies. This scheme is designed for big corporations and companies with a large number of employees.

An industry scheme is a scheme specially established for industries with high labor mobility, for example, the catering and construction industries. An employee does not need to change the scheme if he or she changes employment within the same industry. When changing employment, the employee has the choice of switching the benefits to the scheme of the new employer or remaining in the current scheme.

Trust Arrangement

All MPF schemes must be established under a trust arrangement that are governed by Hong Kong law. Scheme assets will be held separately from the assets of the trustees or the investment managers. This safeguards the interests of the scheme members from unnecessary financial risks.

Mandatory contributions are basically calculated on the basis of 10% of an employee's relevant income, with the employer and employee each paying 5%. Self-employed persons also have to contribute 5% of their relevant income. Mandatory contributions must be paid to registered MPF schemes managed by trustees. They must be paid for each period for which an employer pays relevant income to his or her employee. Investment managers will be appointed by the trustees of MPF schemes to make long-term investment of scheme assets and accrue benefits for the scheme members for their retire-

Figure 2: MPFA Scheme

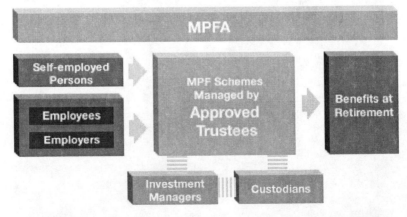

ment protection. The diagram in Figure 2 shows the structure of the MPF arrangement and the interrelation of the various bodies.

The accrued benefits of an employee can be transferred to another scheme when the employee changes jobs. The MPF legislation has prescribed 65 as the retirement age. Scheme members who have attained age 65 may withdraw the benefits accrued from mandatory contributions in their MPF schemes in the form of a lump sum.

Scope of MPF Business

MPF-related services represent a potentially huge market. Before the MPF launch, less than one-third of the workforce of 3.4 million people already had some form of retirement protection. Those are exempted from participation in the MPF scheme. A further 400,000 do not qualify for inclusion for various reasons. The MPF scheme is targeting the remaining workforce. The Government has estimated that annual MPF contributions would amount to more than HK$10 billion (approximately US$1.3 billion) in the initial years, increasing to about HK$40 billion (~US$5 billion) when the system matures. However, this assumes that MPF service providers will have to be able to retain existing customers who join initially lest subsequent erosion in sales may occur. Increasing sales from the MPF business provides an ample opportunity for cross-selling, mainly other financial products, allowing the client base to broaden and business to expand. There is no doubt that a market transformation process has begun that will significantly change the way retirement protection services are used in Hong Kong.

The Playing Field

The MPF market is presently heavily regulated by the Mandatory Provident Fund Authority and the Security and Future Commission with detailed and specific responsibilities imposed on the trustee as well as the investment manager. The nature and business of providing MPF service and the high compliance requirements of the regulatory bodies has limited the entry of players. However, the banks, fund management houses and insurance companies are now all preparing to expand their business into the MPF market. Based on their own estimations, major players totally claimed 285% of the market share (Table 1).

Table 1: Market Share

Financial Institution Sectors	Claimed Market Share
Major Banks	120%
Major Insurance Companies	145%
Other Fund Management Houses	20%
TOTAL	285%

Obviously, this means that most players are overestimating their market share or market potential. However, the figures do reflect the relative market shares among the different sectors, with banks and insurance companies dominating the MPF market.

The Mandatory Provident Fund Schemes Ordinance was enacted in 1995 to set up and regulate the MPF system. MPF is a mandatory, fully funded contribution scheme that is privately managed. The Hong Kong Government's roles are mainly as policy maker, regulator and supervision body of the MPF market. The decision to put fund management in private hands and to create a competitive market environment was based on the belief that a market-based model would be more cost-effective and would ultimately benefit the retirees. At the end of 2000, there were 20 approved trustees and 31,927 registered MPF intermediaries competing to provide fund management services. Table 2 shows that despite the expiration of the legal deadline, about 20% of the workforce and 30% of the employers had not enrolled in MPF.

SETTING THE STAGE

Business strategies may be profit-oriented or may emphasize market share. The latter appear currently as the predominant strategy for most providers. In the beginning, there has been some uncertainty as to when employers will actually sign up for the MPF scheme. Many clients maintained a wait-and-see attitude while they tried to learn more about the offered plans and the providers before actually committing to invest in their preferred plan. Service providers have used this time window to improve their competitive

Table 2: Some Crucial MPF Statistics (as of February 15, 2001)

	Number
Approved Trustees	20
Registered MPF Intermediaries	31,927
Corporate Intermediaries	463
Individual Intermediaries	31,464
Enrollment	
Employers	70%
Employees	80%
Approved MPF products	243

position in the market (Clemons, 1986). For example, some have improved their e-commerce systems and user interfaces in order to make their e-MPF services more attractive to clients, others have launched extensive promotion campaigns to educate the market about their services.

Intense competition exists among the MPF providers for supplying a convenient one-stop-shopping or outsourcing service. Most service providers view the MPF service as a complementary product within their entire product portfolio. For example, banks are increasingly competing on their e-banking capabilities and see e-MPF as an attractive new product to service their customers. Banks and insurance companies alike are developing customized, comprehensive packages that allow customers one-stop shopping for all financial service needs.

However, there are various costs such as those associated with the initial set up of a scheme, annual trustee and fund management services fee, switching fee, buy-and-sell spread, cash rebate, discount, associated benefit packages, payroll/MPF contribution conversion software as well as many others. In addition to after-sales service, it is a combination of some or all of the above that provides the appeal to the employer when making an MPF selection decisions.

The MPF market is dominated by big players with large promotion budgets that aggressively compete for new customers. Advertising campaigns typically elaborate on the theme that MPF is an investment for life, but they also emphasize the importance of the financial stability and security of the provider. Market leaders like HSBC and the Bank of China (BOC) currently have the edge. Distribution through retail bank customers is still a reliable tool. Appendix A shows the number of branches/agents in addition to the claimed market share.

The banking sector and the insurance sector hold about one-half of the MPF market share each, according to Table 1. The MPF provider has to incorporate expertise in trustee services, fund administration, custodial service and investment management. The banks have a distinct advantage over the other types of investors. Some banks, like HSBC and BOC, have developed their own integrated e-MPF platform. Others, like the Standard Chartered Bank, have decided to outsource some or most of the services. The wide range of financial products offered by the banks make this sector a unique battlefield. Traditional banking services, credit card business, mortgage, payroll, commercial loan and documentary credit facilities are products that can be enriched and value added if complemented with MPF services. However, they need to employ all their resources, as the business potential of cross-selling other types of banking products is substantial.

Failure will ultimately mean losing a valuable customer and not just his or her MPF business.

Manulife is currently the only insurance company that has associated business units that can provide one-stop-shopping service. Some other MPF providers from the insurance or fund management background have a strong client base but lack expertise in some areas of the MPF service. Hence, forming alliances with other financial institutions is the preferred strategy of companies like, for example, CEF Life, CMG Asia or Principal Trust. Business is generated through their agents or other sales persons. Their traditional products are the various commercial and personal insurance products.

The competitors in the MPF market try to differentiate their services based on their fee schedules and, increasingly, on other product differentiators such as customer service and personalization (Hanson, 2000, pp. 151-219). At the early stage of introducing MPF services to the market, most providers view customer acquisition as the main business goal. Marketing strategies emphasize low fees to attract new customers (see Appendix B for the comparison of schedule fees). HSBC was the first to offer fee waivers and bonus fund units. Manulife and AIA offered cash bonus while BOC and CEF offered fees as low as 1.5%. It is expected that the price war will continue and intensify for some time, with companies offering heavy rebates, discounts and fee waivers to attract and retain the customers and employers. But there are provisions in the participation contracts that allow service providers to increase fees in the future. Customers, however, demand more than just low fees when choosing a particular scheme. As a response, providers have begun to differentiate their service products through other criteria such as customer service, e-service friendliness and plan flexibility.

In the longer run, the fees charged will increase as all providers have similar provision in the participation agreement to allow fee adjustment. Hence, the ability for MPF service providers to provide a better customer service becomes an important means for both customer acquisition and retention. Leading banks are arranging MPF workshops for each employee in addition to call centers where representatives are on duty answering questions up to 16 hours a day. Interactive voice response systems help provide information to customers, and Wweb-based services now offer electronic document transfer as well as fund switching. All providers now routinely offer these basic services.

Providers also offer an assortment of bundled benefits such as waiving the annual fee for credit cards, offering preferential commercial or mortgage loan interest rates, free online banking membership, lower insurance premiums, better deposit rates, convenient payroll arrangements, gift coupons, cash rebates, souvenirs and other bonuses. Appendix C shows the non-fee competition offered by the key players.

CASE DESCRIPTION

We now introduce the e-business model underlying the MPF service in Hong Kong's public e-environment. Based on an analysis of the e-business model of one of the leading MPF service providers, we discuss how e-commerce technologies can be applied to achieve success in the competitive MPF market and provide benefit to the various players. Our case subject is a leading bank that engages in a wide range of businesses such as traditional banking, unit trusts, asset management, institutional, corporate and private client portfolio, trustee services, custodian services, as well as retirement fund services. The bank has over 200 branches and has been providing a full spectrum of banking services for a long time in Hong Kong. It has been managing hundreds of retirement fund schemes and billions of dollars in assets. The bank has especially targeted small and medium-size employers, that is, companies with 5 to 1,000 employees. It aims at gaining 15% of the market share, managing $4.8 billion annual fund inflow from 48,000 employer schemes and 480,000 employee accounts. It has launched three different investment schemes.

In the first phase of the marketing activities, MPF products have been sold through the existing clients of the bank. Later, this channel will expand to other potential targets including the valued customers of the other banking or insurance sectors. It will be up against both fee and non-fee competition and is more than aware of the importance of using technology to bring about a competitive advantage. Using this particular MPF service provider as the base case for this analysis, we examine the design of the main components of the e-service model, and discuss how e-commerce technologies facilitate B2E/B2C, B2B and B2G transactions (Timmers, 1998).

In the traditional business model of the old, physical economy, wealth is correlated to physical assets and companies go vertical for economies of scale. There is noticeable scalability that is related to company size and tangible assets. Market share and profits benefit from proximity of production, sales and distribution. It is a sellers' market with information arbitrage. In the new, digital economy, on the other hand, wealth is correlated more strongly with IP assets, knowledge and customer relationships. The new model emphasizes the creation of value networks (Tapscott, Ticoll & Lowy, 2001). Scalability is related to the ability to encapsulate and leverage knowledge. Virtual channels and digitization remove barriers of time, place and scale.

The diagram in Figure 3 highlights the changes from the old to the new economy.

As Figure 4 indicates, the economy continues to transform. The suppliers are no longer at the center of business activities as the buyers instead take the leading role, enabling a more mature e-market and consumer-centered economy.

Figure 3: The Change of E-Business World

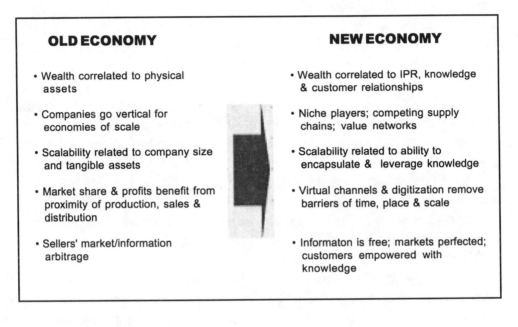

OLD ECONOMY

- Wealth correlated to physical assets

- Companies go vertical for economies of scale

- Scalability related to company size and tangible assets

- Market share & profits benefit from proximity of production, sales & distribution

- Sellers' market/information arbitrage

NEW ECONOMY

- Wealth correlated to IPR, knowledge & customer relationships

- Niche players; competing supply chains; value networks

- Scalability related to ability to encapsulate & leverage knowledge

- Virtual channels & digitization remove barriers of time, place & scale

- Informaton is free; markets perfected; customers empowered with knowledge

Figure 4: Transformation of the Economy

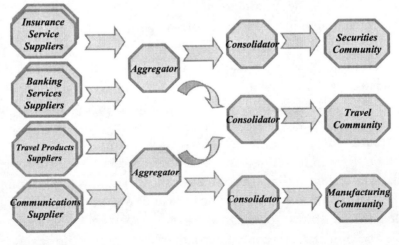

Source: FMM

E-Service Benefits of MPF

Electronically supported Mandatory Provident Fund management (e-MPF) is a good example of the new business possibilities in the information age. We can identify several properties of the e-MPF business model, depicted in Figure 5, that provide benefit over conventional approaches:

Electronic Money: All monetary transactions are based on electronic money transfers.

Digital Distribution: In the old model, distribution of paper-based documents accounted for more than 30% of the total service cost. In the e-MPF model, all transactions and instructions are made in electronic form, which substantially reduces distribution cost.

Knowledge and Information: In order to retain their most profitable customers, MPF customers should be provided with the most accurate information. Customers are given access to a wealth of information that allows them to make their own decisions. More flexible and customized investment plans can be offered.

Other Complementary Products: The banks, insurance companies and investment institutions are the three main providers in the MPF service. Their current products can be totally transformed into electronic formats and integrated with MPF products. The service providers can offer their entire product portfolio through the same transaction platform.

Re-Intermediation: Service providers no longer specialize in just one part of the value chain. They become aggregators who provide any financial service to their customers in connection with MPF. We observe a move from value chain management to value network management.

Personalization: New e-commerce technologies like WebBots and data mining enable the provision of cost-effective personalized information and transaction services at large scale. Several types of customer interactions can be automated, for example, contribution statement inquiries, plan switching, change of contribution percentage, change of employer or change of trustee.

Cross-Selling: Customer databases can be employed to automatically track transaction record and generate personalized recommendations for other service products that might be relevant and interesting to the customer.

The challenge of a successful e-MPF implementation is to develop a transparent, seamless alliance-based value network (Bakos & Brynjolfsson, 1998) that allows the customer to choose the best product combination for his or her personal needs without having to research the complex supplier market. Convenience and flexibility must be provided without compromising on security and privacy issues. Since e-MPF services have just been launched, it is too early to tell whether or not these challenges can be met. It will probably take some years before a reasonable assessment can be made.

Business-to-Consumer (B2C) and Business-to-Employee (B2E) Transactions

Account management and changes in chosen investment plans are very labor intensive in the conventional MPF model. Switching to the e-MPF model promises to significantly lower administrative cost while, at the same time, increases flexibility. As a contributor to an MPF scheme, each individual employee has an interest in checking how his or her contributions are invested and how much has accumulated in their MPF account. Providers have set up Websites that provide personalized MPF account management and disseminate updated MPF information and investment decision aids.

Customers are able to access or download information selectively. Employing data mining tools and Web usage tracking technologies based on page hits and service requests (Choi & Whinston, 2000), providers learn about their customers' interests. This allows them to respond more quickly to problems and newly arising demands. Expensive human resources can be shifted from sales and clerical activities to customer service.

Low operability and controlability in the traditional model gives the employees a sense of MPF basically being a salary deduction rather than investment for retirement protection. With a Website capable of integrating

Figure 5: E-MPF Business Model

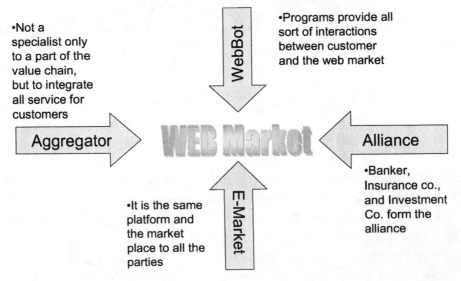

E-MPF Business Model

- •Not a specialist only to a part of the value chain, but to integrate all service for customers

WebBot

- •Programs provide all sort of interactions between customer and the web market

Aggregator → WEB Market ← **Alliance**

- •It is the same platform and the market place to all the parties

E-Market

- •Banker, Insurance co., and Investment Co. form the alliance the MPF system, employees are able to query the performance of their contributions online. This is especially critical when the financial markets show changes in volatility. For example, during a slump in the stock market, employees may want to change their schemes from high growth to stable growth in order to prevent or reduce losses.

The e-MPF model enables the transformation of the service process that adds value to providers and clients alike. For example, plan or scheme switching can be streamlined. Although it is an agreement between MPF providers and employees, in many cases, the employers act as middlemen to collect the data and input them into their payroll system and then notify the providers. This process incurs a high administrative overhead in the conventional model. In the e-MPF model, on the other hand, an automated MPF plan switching workflow can be setup on the Web to reduce administrative costs incurred by both MPF providers and employers. During month end, employers can print the change in employee plans, or download it from the Web and upload it into their payroll systems. New plan switching can be effective immediately upon employees' data input, or be effective after employers' confirmation.

Unlike the existing retirement schemes, the governing MPF authority body (MPFA) poses a legal requirement on MPF of a 5% minimum employer contribution and a 5% minimum employee contribution. However, MPF allows the employer or employee to contribute more than 5% voluntarily. However, changing the contribution percentage involves a tedious workflow among three parties namely the employee, employer and MPF providers. In the traditional MPF model, the employee or employer states his or her intention to change the employee or employer contribution percentage respectively. Then the employer, as a middleman, collects the changes, inputs data into the payroll system, notifies the bank to transfer correct funds to both employee and MPF, and finally notifies the MPF provider of the changes. The MPF providers will then credit the corresponding contributor's account upon the fund transfer.

In the e-MPF model, however, contribution percentage changes can be supported through a Web-based workflow system. Both employee and employers can change the employee or employer contribution percentage over the Web. Every month, employers can download the revised contribution percentage, confirm it and upload it into their payroll systems.

Web-based MPF service models can provide passive cross-selling through referral links to other bank products such as investment options, insurance products or online banking. A more proactive way is to adopt rule-based cross-selling. For example, if the contributor has a salary increase of more

Figure 6: Workflow of Change in Contribution %

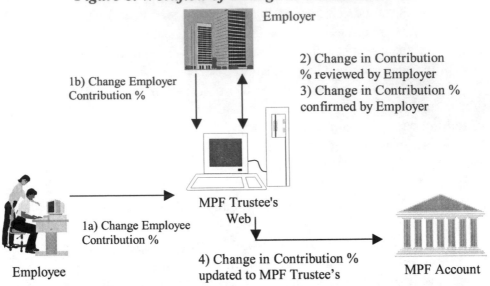

than 10%, it would indicate that he or she is likely to be more interested in a cross-sell of other investment products. If a contributor indicates that they wish to change the plan from high growth to stable, they are adopting a more conservative approach. The website can also be used to actively cross-sell other products.

Rule-based software agents can perform cross-selling functions in two ways. First, it can track and trace Website access of customers, and activate cross-selling and/or up-selling offers if certain rules are matched. For example, if an employee is going to change to a high-risk plan, we can proceed to cross-sell some investment product. In alignment with appropriate marketing strategy, it is possible to use it to devise data-signaling programs, or data-screening programs for product customization. Second, employees could be allowed to set rule-based e-mail notification. For example, he or she could set a rule to receive e-mail, if the return of their plan is lower than 10%, or if the high-risk plan is gaining a return of over 30%. Also, the employee can choose to receive customized e-mail notifications on new investment products, insurance products or financial news. Web-based personalization methods can provide convenience and help strengthen the customer relationships.

The Web can also be used to provide other value-added services to the contributors, for example, investment analysis tools. This function can be used to primarily trace and track the contributor's expected risk and return and monitor the markets or stocks they prefer. Software agents can act as personal investment advisors and assist contributors in their investment decisions.

To achieve a reasonable level of data and transaction security, e-MPF models employ standard security management technology. Secure Socket Layers (SSLs) are adopted to provide authentication and data integrity. The users can simply login with an encrypted user ID and password. This ensures that the integrity of data transmitted from MPF providers, for example the MPF balance query, will be maintained. In addition, the secure electronic transfer protocol (SET) and digital signatures can be used to give enhanced security on eavesdropping and non-repudiation. However, this is currently not used much in MPF transactions at this stage because using SET and digital signatures require that the participants have obtained a personal digital certificate, which is currently rare in Hong Kong.

The introduction of the B2C and B2E e-commerce features discussed above are expected to bring the following benefits to both employers and employees:

Updated MPF Information: The first benefit is online, updated MPF information including MPFA regulation, MPF products, FAQ of MPF as well as links to other bank products. This is a unique feature that cannot be offered by other channels of communication such as television or newspaper. In addition, it provides another medium to perform pre-sales advertisement and collect feedback from employers and employees.

MPF Service Websites: Secondly, MPF service Websites can help to provide customers with a comprehensive and convenient one-stop-shopping place for a wide range of financial services. The higher switching cost of moving the entire service package to other providers may increase customer loyalty. Increased flexibility of MPF account management should increase customer satisfaction.

Customer Relationship Management: As more employment and financial details can be collected from the Web, it helps to build better customer relationship management. Consumer databases help providers to better respond to customer needs. Providers can cost-effectively provide higher degrees of customization and personalization, and thus offer better products to their customers.

MPF Service Platform: The MPF service platform can be integrated with e-banking and other e-services, and can be used for cross-selling and promotions.

Business-to-Business (B2B) Transactions

First, we discuss the business-to-business processes between the trustee and the company that employs individual MPF participants. After signing an agreement to join the MPF service provided by the trustee, the employer needs to send information about every employee to the trustee for creation of a new MPF account. Information about the percentage of the monthly contributions—a minimum of 5% of the salary, the particular scheme that the individual has selected, personal data and other transaction-oriented information—has to be sent to the trustee. When these initial stages are completed, the company is required to send payroll information to the trustee regularly so that contributions to the MPF funding can be arranged. After the monthly contributions from the employees are received, the trustee needs to send a receipt to the company. Meanwhile, if a new employee is hired or an employee leaves the company, updated employee information needs to be transferred to the trustee. At the end of each year, the trustee is required to generate an annual report about the company's contributions to the MPF schemes of its employees.

To sum up, high volume document exchanges are required between the trustee and the employer on a regular basis. All these exchanges are concerned with the transfer of information which suggest to automate this process using modern e-commerce technologies.

Employee Information Transfer: Information about the employee can be transferred from the company to the trustee via the Internet. After setting up the employee's individual MPF account, the company can use the Internet to transfer and update information, including information about new employment or termination of employment contracts.

Payroll Information Transfer: The company can send monthly payroll information about the each employee to the trustee in an electronic format. This information can automatically trigger the transaction of MPF contribution from the employer account to each employee's MPF account. To ensure security about the transaction, all details about to be transferred have to be predefined clearly before the transaction.

Confirmation Information About Contribution Transaction: After each successful transferral of the contribution from the employer's account to the employee's MPF account, a confirmation message is sent to the employer via the Internet to complete the process. These messages serve as formal receipts from the trustee. Annual reports about the employer's contributions are also transferred from the trustee to the employer over the Internet. As non-repudiation is required for the process, SSL is typically used as an encryption method.

Balance and Analysis: With all the above processes automated by electronic means, it is now possible for the employer to have online access to the balance of each employee account. In addition, report and real-time analysis about the investment of the MPF funding can also be offered over the Internet.

In the case that an employee leaves his company and joins a new employer with a different MPF trustee, the original employer should send updated employee information to its own trustee to terminate the contribution to the employee's MPF account. When the employee wishes to join the MPF scheme of a new trustee of the new company, he can send his request to the original trustee to ask for transmission of past MPF information to the new trustee, if the new trustee has also adopted an e-MPF model. In this case, switching employer and trustee is completely automated.

To provide a standard way for information exchange over the Internet, XML has been suggested as the most effective way to exchange standard MPF documents (Choi & Whinston, 2000, pp. 45-50). With XML, we can define information in a document by tags, that are predefined by the trustee, so that information can be automatically processed and trigger content-based actions. Figure 7 shows a simplified example of an XML-tagged MPF document.

Most of the MPF-related information exchanged between employers and trustees is confidential as it involves personal particulars as well as payroll information. Therefore, security concerns have to be resolved. Measures have to be taken to ensure confidentially, authenticity and non-repudiation of information being transferred. All data transmitted over the Internet are encrypted using SSL. In Hong Kong, the public key infrastructure that is required for SSL encryption has been established by the Hong Kong Post Office.

The transfer of the contribution between accounts needs a higher level of security as it involves transfer of money. Whenever a contribution is made, money has to be transferred from the employer's account to the employee's MPF account. All money transactions need to be predefined before the actual transaction occurs (e.g., from what account to what account, the upper limit

Figure 7: XML-Tagged MPF Document

```
<?xml version = "1.0" standalone ="yes">< 
<transactionconfirm>
<trustee>MPFBank</trustee>
<company>ABC manufacturer</company>
<date>2000/04/30</date>
<periodstart>2000/04/01</periodstart>
<periodend>2000/04/30</periodend>
<contribution>HK$400,000</contribution>
<status>successful</status>
......... .
</trsansactionconfirm>
```

of each transfer, the frequency of the transaction, etc.). The real transaction from the employee's account to the MPF account is done through the bank's secure transaction system instead of over the public Internet. The Internet only serves as the medium for the company to send payroll information to trigger the predefined transaction. In this way, money transfers are all done through the banking system's own computer network. The Internet only serves as the media to trigger predefined transactions securely. If the employer account and the MPF account are from the same bank, the process is much simpler.

After each transaction, the database is updated immediately so that the employer can verify the status of all transactions over the Internet.

The whole process can be summarized in the diagram that is shown in Figure 8.

Several benefits are expected from the automation of the B2B process in the MPF service schemes:

Reduction of Administration Cost: Without the electronic process, the cost of exchanging information between the company and the trustee involves lots of manual work, use of paper documents, delivery charges and time delay due to delivery time. Using electronic information exchange, all these processes can be done speedier and at much lower cost.

Error Reduction: Without automation, data needs to be entered and reentered at several stages in the process. This inevitably introduces clerical error. With electronic data exchange, the data is only entered once and reused in all subsequent steps. Thus, human error in data entry can be significantly reduced.

Figure 8: Summary of MPF Process

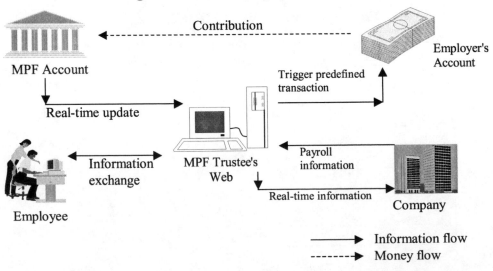

Fast Response: Online transactions are executed almost instantaneously, which dramatically reduces cycle times and helps streamline the entire process.

Business-to-Government (B2G) Transactions

Employers have an obligation to report to the MPFA, the government body that supervises the MPF industry, about its progress when joining the MPF scheme, and inform about changes and provide updates on their regular MPF activities. They also have an obligation to send documents to the MPFA to prove that they have really contributed to their employee's MPF account. Trustees also need to send documents and reports about their investment policy, financial balance and investment details to the MPFA for inspection. This is required for supervision of trustees on their investment with the funding. The documents required for the MPFA are mostly legal documents and financial reports that include the following: employer application form, principle brochure, employer sales booklet, participation agreement, trust deed, proof of employer's contribution, MPF balance and MPF investment report.

As the processes between the employers, the trustees and the Government involve a large volume of information transfer, it is also very suitable to use the Internet as the platform to perform all these tasks (Nelson, 1998). The opportunity of B2G electronic process currently exists but it requires action on the part of the Government to initiate the process. Since the Hong Kong Government has recently, at the dawn of the 21st century, publicized its 'Digital 21' initiative, which is aimed at promoting digital technologies to help transforming Hong Kong to a knowledge-based society, it can be expected that the e-MPF model will soon be extended to include the implementation of the B2G processes. The Hong Kong Government has already launched a series of electronic government services in other areas, and it is expected that a full e-MPF implementation, including the B2G part, will follow soon.

Current Challenges/Problems Facing the Organization

Government and industry have pushed the introduction of the MPF system. Only a few years after the initial decision, the MPF is up and running. This is certainly a success. As discussed in the previous sections, the MPF model is conceptually sound and the e-MPF model, in particular, promises great benefits to everyone. However, several serious problems emerged during the implementation, which we will summarize and discuss below. While some problem areas are concerned with technical issues, the most serious difficulties are related to managerial problems. Questions arise whether some of these problems could have been anticipated and better managed.

Challenges arise from the multiple levels of interactions between the parties involved. What are the fundamental features and components that would lead to a successful e-service model (Timmers, 1998)? The e-MPF model needs to include four sets of carefully designed relationships between the main parties: relationships between employers and service providers (business-to-business, or B2B relationships); service providers and clients (business-to-consumer, or B2C relationships); employers and employees (business-to-employee, or B2E relationships); and employers and service providers with government (business-to-government, or B2G relationships). We have introduced the e-MPF design and discussed how it implements the main stakeholder relationships. Complex information services involving multiple parties require a high degree of coordination to avoid friction and conflicts. Complexity arises from multiplicity of stakeholders, coordination requirements, new types of large-scale services and the different roles played by all parties involved. This results in a set of service implementation challenges. What are the main technical difficulties when implementing such a complex information service? And last, but not least, what are the main managerial issues in developing and launching a new service like MPF? What problems have occurred and what has their impact been? Who is responsible? Could some of the problems have been anticipated and perhaps avoided? Five areas have been identified as the main problem sources.

1) *Market Potential*: Many companies, especially small, family owned enterprises and self-employed professionals, did not fully cooperate in implementing the MPF model. As of May 2001, about 30% of the businesses and 10% of the employees have still not enrolled, although the enrollment deadline has long passed. The government is facing a big problem of how to deal with thousands of companies and employers violating MPF regulations. Prosecution has been threatened but mass prosecutions on that scale appears unenforceable. Critics have claimed that many small businesses could simply not afford the mandatory MPF contributions. Enforcing enrollment could then lead to layoffs. Some legislators have already blamed the MPF scheme for the recent rise in unemployment, which has gone up to 4.5% in February 2001, the first increase in 18 months. Many low salary workers who already suffer financial hardship cannot afford a 5% reduction of their net pay. The image of the MPF has suffered at large from negative press. The slowdown of the U.S. economy and volatile stock markets have nega-

tively impacted the initial performance of many MPF funds. More than half of the 290 MPF investment funds had posted negative returns by April 2001. Overall, the MPF has become quite unpopular, and most people only participate because law requires it.

2) *E-Business Limitations*: The e-MPF market is currently only a small fraction of the total MPF market. Many employers simply follow standard plans chosen by their company as default options, and do not use the flexible e-MPF mode. Despite the high Internet penetration rate in HK, there remains a large portion of the workforce that doesn't use computers at all or doesn't use sophisticated online services like the e-MPF. And even among the Internet-savvy professionals, e-MPF adoption has been relatively slow (Bhatnagar, Mistra & Rao, 2000).

3) *System Integration*: Multiple parties are involved in the implementation of the MPF system—the MPFA (Mandatory Provident Fund Scheme Authority) who enforces the law, the trustees who manage the funds, the intermediaries who provide daily administrative work and user interface, and the client companies who provide payroll data. All those organizations developed their own systems. A seamless integration of all business-to-business processes is an absolute requirement for e-MPF to work successfully. During the implementation phase, however, there were numerous coordination problems among the involved parties that delayed projects and led occasionally to ad-hoc style patchwork to fix system flaws. Back-end integration with payroll legacy systems at the employer's side with the MPF system is another important issue. There are very different legacy systems in place, running on different computing platforms. Many companies have had difficulties with the integration of their legacy systems. Similarly, service providers have encountered problems related to the integration of the e-MPF service with their other e-services. Each service provider has developed its own e-MPF system, which has also led to a proliferation of different user interfaces. Better coordination between the providers could have reduced or eliminated this problem.

4) *Project Management*: During the software implementation phase of the MPF system, most service providers were running behind schedule and over budget. A large number of the MPF software implementations were outsourced to international software vendors and consultants. When the management realized that their projects were off target, it was often difficult to regain control, especially when the projects were outsourced. Only few clients from the employer's side enrolled in the early development stages so that the service providers could not really familiarize

themselves with the client's systems. This led to some of the integration problems cited above. As deadlines approached, managing the relationships between banks and the contractors became increasingly challenging; extra resources had to be committed to meet the deadlines. There was little time for system testing. Testing with real data and cases was impossible because most companies had not enrolled until the very last moment before the legal deadline. System integration tests were impossible because the other parties were not ready yet. When the MPF systems were rolled out, most systems were not fully tested and there was a high risk of major problems. Fortunately, no disastrous failures have been reported, which many commentators attribute more to luck than to proper problem management.

5) *Flexibility:* At the moment, most e-MPF systems allow users only to adjust their investment plans one to several times a year, with only three kinds of investment choices, ranging from low risk to high risk. The industry is currently not fully utilizing the e-MPF business model, which may deter employees from joining the e-service version of the MPF. At the same time, service providers postpone system upgrades and improvement, citing the currently low usage level of their e-services.

A final assessment of the success or failure of Hong Kong's MPF initiative, especially in terms of MPF as an electronic service, will have to wait a few more years until MPF has matured. An interesting future research project could be a longitudinal study that investigates how the MPF market matures and how the players respond to the various challenges.

APPENDIX A

Major Competitors

Table A: Banking

	MPF Organization Mode	Integral Parts of MPF Services	Number of Branches	Market Share (Claimed)
HSBC/ Hang Seng Bank	A separate business division	One-stop-shop	360+	40%
BOC	Joint venture with Prudential Assurance	One-stop-shop	270+	15%
Bank of East Asia	A separate business unit	One-stop-shop	100+	5-15%
Bank Consortium	Alliance with banking entity	Outsourcing Fund Management	280+	10-15%
Standard Chartered Bank	A separate business division	Outsourcing Fund Management, Trustee and Administration	80+	15-20%
TOTAL				*100-105%*

Table B: Insurance and Fund Management

	MPF Organization Mode	Integral Parts of MPF Services	Number of Agents	Market Share (Claimed)
AXA	A separate business unit	Outsourcing Custodian Service and Fund Management	3,000	20%
CMG	Partner up with Butterfield Trust	Outsourcing Custodian Service	2.500	20%
AIA/Jardine Fleming	Joint venture	One-stop-shop	8,000	25-30%
Manulife	A separate business unit	One-stop-shop	2,300	20%
INVESCO	Handles Fund Management only. Partner up with Bermuda Trust.	Outsourcing Trustee, Custodial & Fund Management	100+	10%
SUB-TOTAL				*90-100%*
OTHERS				*25-45%*

APPENDIX B

Table 1: The Fee Schedule (The Initial Offer)

	Initial Charges	Annual Management Fee-capital preservation fund	Annual Management fee - other funds	Special Feature
HSBC/ Hang Seng Bank	1.0%	2.200%	1.9500%	Clearest definition of the various types of fees
BOC	1.5%	0.800%	1.4875%	Cheapest preservation fund
Bank of East Asia	1.0-1.2%	1.200%	1.5000%	——
Bank Consortium	2.0%	1.400%	1.6250%	Least choice of available fund
Standard Chartered Bank	2.5%	1.675%	1.8000%	Most choice of available fund
AXA	1.0%	1.0000%	22.000%	Converting deposit into subscription is possible.
CMG	3.0%	2.500%	2.5000%	Lump sum fees available.
AIA/Jardine	0.5%	2.000%	2.0000%	Unlimited number of switch of fund over the net and phone

APPENDIX C

Table 1: Non-Fee Competition-Technology and Bundled Benefit

	MPF specific Website	*HRIS*	*IVRS*	*Associated Banking/Insurance Bundled Benefit*
HSBC/ Hang Seng Bank	Yes www.mpfdirect.com	Payroll / MPF Packages	IVRS	Yes
BOC	No www.bocgroup.com	Payroll & MPF Adaptor	IVRS & MPF Workshop, Call center	Yes
Bank of East Asia	www.hkbea.com	Payroll & MPF Adaptor	IVRS	Yes
Bank Consortium	Yes www.bcthk.com	Payroll software	IVRS, investor education	Yes
Standard Chartered Bank	No www.standardchartered.com.hk	Payroll interface system	IVRS	Yes
AXA	No www.axa-chinaregion.com	TBA	IVRS, Call center	Yes
CMG	(TBA)	Payroll interface	IVRS	Yes
AIA/Jardine	Yes www.mpf-aiajf.com	Payroll system	--	Yes
Manulife	Yes www.mpf.com.hk	Contribution Software	Employee Seminar	Yes
INVESCO	No www.invesco.com.hk	Payroll Solution	IVRS, Employee Education, seminar, retirement investment center, MPF news update	Yes

ACKNOWLEDGMENTS

We would like to thank Sarah Cook for helping to prepare the manuscript of this case. We also express our thanks to Simon Leung and Philip Lee for contributions with the MPF market research. Our deepest gratitude, however, has to be extended to Norman Law who provided us with invaluable insights from the banking industry regarding the implementation of the MPF system.

FURTHER READING

Castells, M. (1996). *The Rise of the Network Society,* 1, Oxford, UK: Blackwell Publishers.

Galliers, R.D., Leidner, D. E. and Baker, B. (Eds.). (1999). *Strategic Information Management, 2nd ed.* Oxford, UK: Butterworth Heinemann.

Gulati, R. and Garino, J. (2000). Get the right mix of bricks & clicks. *Harvard Business Review*, May-June, 107-113.

Kalakota, R. and Whinston, A. B. (Eds.). (1997). *Readings in Electronic Commerce.* Reading, MA: Addison-Wesley.

Malone, T. W., Yates, J. and Benjamin, R. (1989). The logic of electronic markets. *Harvard Business Review,* May-June, 166-170.

Munro, A. J., Höök, K. and Benyon, D. (Eds.). (1999). *Social Navigation of Information Spaces.* London, Heidelberg, New York: Springer.

Negroponte, N. (1995). *Being Digital.* London, UK: Hodder and Stoughton.

Porter, M. (2001). Strategy and the Internet. *Harvard Business Review,* March, 63-78.

Strauss, J. and Frost, R. (2001) *E-Marketing,* 2nd edition. Upper Saddle River, NJ: Prentice Hall.

REFERENCES

Bakos, Y. (1998). The emerging role of electronic marketplaces on the Internet. *Communications of the ACM,* 41(8), 35-42.

Bakos, Y. and Brynjolfsson, E. (1998). Organizational partnerships and the virtual corporation. In Kemerer, C. F. (Ed.), *Information Technology and Industrial Competitiveness: How IT Shapes Competition,* 49-66, Kluwer Academic Publishers.

Bhatnagar, A., Misra, S. and Rao, R. H. (2000). On risk convenience and Internet shopping behavior. *Communications of the ACM,* 43(11), 98-105.

Clemons, E. K. (1986). Information systems for sustainable competitve advantage. *Information and Mangement,* 11, 131-136.

MPFA: Mandatory Provident Fund Schemes Authority. (2001). The Official MPF Web Site. Retrieved on the World Wide Web: http://www.mpfahk.org.

Nelson, M. R. (1998). Government and governance in the networked world. In Tapscott, D., Lowy, A. and Ticoll, D. (Eds.), *Blueprint to the Digital Economy,* 339-354. New York, NY: McGraw Hill.

Soon-Jong Choi, S. J. and Whinston, A. B. (2000). *The Internet Economy: Technology and Practice,* Austin, TX: Smart Econ Press.

Tapscott, D., Ticoll, D. and Lowy, A. (2001). *Digital Capital: Harnessing the Benefit of Business Webs,* Harvard Business School Press.

Timmers, P. (1998). Business models for electronic markets. *EM-Electronic Markets,* 8(2), 3-8.

Ward Hanson, W. (2000). *Principles of Internet Marketing,* Cincinatti, OH: Thomson Learning.

Westland, C. J. and Clark, T. H. K. (1999). *Global Electronic Commerce: Theory and Case Studies,* Cambridge, MA: MIT Press.

BIOGRAPHICAL SKETCHES

Theodore H.K. Clark has taught information technology courses at the Hong Kong University of Science & Technology's (HKUST) Business School, Harvard University and Wharton Business School (University of Pennsylvania). He is an Associate Professor in Information Systems and serves currently as the Deputy Head of the ISMT Department at HKUST. His current research interests included electronic commerce, information technology (IT) strategy, business process reengineering, supply chain management and information infrastructure development. Dr. Clark has published a number of articles in leading research journals within the IT research community and has also written many case studies published by HKUST and HBS on electronic commerce, IT strategy and business process reengineering. In addition, he is the co-author of a book entitled Global Electronic Commerce: Theory and Cases, *published by MIT Press.*

Karl Reiner Lang is an Assistant Professor in Information Systems at the Hong Kong University of Science & Technology (HKUST). Before joining HKUST in 1995, he had been on the faculty of the Business School at the Free University of Berlin. Dr. Lang has been teaching courses on Information Technology and Electronic Commerce at the undergraduate, postgraduate and executive levels. His research interests include management of digital businesses, decision technologies, knowledge-based products and services, and issues related to the newly arising informational society. Dr. Lang's recent publications have appeared in leading research journals such as Annals of Operations Research, Computational Economics, Journal of Organizational Computing and Electronic Commerce, *and* Decision Support Systems. *He has professional experience in Germany, the USA, and Hong Kong.*

Will Wai-Kit Ma is a Research Associate in the ISMT Department at the Hong Kong University of Science & Technology. Will had worked as Marketing Manager in a multimedia company coordinating multimedia projects and application training courses for five years. He then became an Instructor training practicing teachers computer applications. He is also the author of a number of popular computer books, including Own Your Business on the Internet, A Guide to Hong Kong Web Sites, A Guide to Internet, *and* CD-ROM Practice Q&A. *His research interests are in information technology adoption, computer attitudes, electronic business and IT in education.*

Enabling Electronic Medicine at Kiwicare: The Case of Video Conferencing Adoption for Psychiatry in New Zealand

Nabeel Al Qirim
Consultant, New Zealand

EXECUTIVE SUMMARY

Telemedicine emerges as a viable solution to New Zealand health providers in reaching out to rural patients, in offering medical services and conducting administrative meetings and training. No research exists about adoption of telemedicine in New Zealand. The purpose of this case study was to explain factors influencing adoption of telemedicine utilizing video conferencing technology (TMVC) within a New Zealand hospital known as KiwiCare. Since TMVC is part of IT, tackling it from within technological innovation literature may assist in providing an insight into its adoption within KiwiCare and into the literature. Findings indicate weak presence of critical assessment into technological innovation factors prior to the adoption decision, thereby leading to its weak utilization. Factors like complexity, compatibility and trialability were not assessed extensively by KiwiCare and would have hindered TMVC adoption. TMVC was mainly assessed according to its relative advantage and to its cost effectiveness along with other facilitating and accelerating factors. This is essential but should be alongside technological and other influencing factors highlighted in the literature.

BACKGROUND

Telemedicine emerges as a viable solution to New Zealand health providers in reaching out to rural patients, in offering medical services and conducting administrative meetings and training. This would improve the cost effectiveness of delivering that service from the standpoint of the institution as well as from the patient's perspective (Wayman, 1994).

All the hospitals in New Zealand are managed by regional organizations known as Health and Hospital Services (HHSs). Some HHSs have one hospital and others have more than one. A survey in the present research found that medical schools in New Zealand were among the early adopters and users of the technology. Out of 23 HHSs in New Zealand, only 12 have actively adopted telemedicine utilizing video conferencing technology (TMVC). The adopted systems ranged between one and four TMVC systems with the majority of HHSs adopting one system only. Those HHSs that adopted one TMVC system use it mostly for general purposes such as managerial meetings, case discussion and occasionally for clinical training. Such initiatives were described as being initial and experimental. Where a HHS owned more than one TMVC system, it was oriented for clinical purposes such as psychiatry, paediatric, dermatology and other medical areas.

Hence, an attempt is made to adopt TMVC to provide prompt and quality medical care even to geographically dispersed patients in different parts of rural areas, which was otherwise not possible or was expensive.

However, despite the rapid growth and high visibility of telemedicine projects in advanced countries like the U.S., relatively few patients are now being seen through telemedicine (Grigsby & Allen, 1997; Perednia & Allen, 1995; Wayman, 1994). A study conducted by Perednia and Allen (1995) found that in almost every telemedicine project in the U.S, tele-consultations accounted for less than 25% of the use of the system. It was mostly used for medical education and administration. The low use of TMVC in clinical activities is violating a principal condition in having it in the first place. The important unresolved issues identified revolve around: (1) how successful the telemedicine can be in providing quality health care at an affordable cost; and (2) whether it is possible to develop a sustainable business model that would maintain profitability over time. This further depends on: (i) clinical expectations, (ii) matching technology to medical needs, (iii) economic factors like reimbursement, (iv) legal concerns (e.g., restrictions of medical practices across state lines, called licensure), (v) social issues (e.g., changing physician behaviour and traditional practices and workflow) and (vi) organizational factors. Wayman

(1994) and Anderson (1997) have endorsed some of the above issues as well. For example, Wayman (1994) pointed to other micro-level implications: many doctors have an aversion to technology; scheduling TMVC encounters with patients represents another burden to clinicians and to technicians; the loss of the one-one-one personal interactions with patients; and patient acceptance to the technology.

The obstacles pointed out by the above results raise concerns about the success of telemedicine as a medical tool. Despite the technological sophistication of the TMVC equipment, it was obvious that its uptake and use specifically in medical areas necessitate addressing organizational, social and environmental factors. Clearly the full potential of the telemedicine technology remains to be realized. Whether this assertion applies to New Zealand HHSs has yet to be identified. Above all, the diminishing funds from the New Zealand government (Neame, 1995) would lead HHSs to consider TMVC as a non-priority tool.

This research was interested in developing an understanding about how one HHS viewed telemedicine and what factors accelerated or hindered its adoption in New Zealand. It specifically focused on the adoption of TMVC in the psychiatry area by one of the HHSs in New Zealand. However, would be quite difficult to emphasize the impacts of the various factors highlighted above on TMVC uptake and use. Therefore it was limited here to technological and to social aspects only. TMVC is a technological innovation, and addressing its technicalities is of paramount importance. Discussing its implications from within a social context would emphasize vital links and interrelationships between the technology and the people adopting it. Accordingly, the role played by the technological innovation theory (discussed in the theoretical framework section) in outlining such implications is well suited here. Other managerial perspectives would be emphasized here in support of the above argument, as it depicts an organizational adoption decision.

For the purpose of this study, the real name of the HHS was suppressed and given the name 'KiwiCare.' Interviews were conducted with main interviewees during the period March-October 1999.

The case study is addressed in the following sequence: review relevant literature about telemedicine; theoretical framework; case background and description; and case analysis and discussion. Towards the end, conclusions are drawn and future areas for research are suggested.

TELEMEDICINE

Telemedicine means medicine from a distance where distant and dispersed patients are brought closer to their medical providers through the means of telecommunication (OTA, 1995; Perednia & Allen, 1995; Wayman, 1994). The value added by this means covers a wide spectrum of benefits through the use of TMVC in consultations, diagnostics, therapeutic, transfer of patient-related records, case management, training and meetings.

Telemedicine was created from the desire to improve utilization of medical resources, specifically scarce education and speciality medicine resources. It has been practised for more than 40 years using various technological means such as telephone, telex and fax (Wayman, 1994).

Some of the technologies used in transmitting clinical data between distant clinical centres were based on telephone calls, fax transmissions, telex messages and VC sessions. Until recently, transmission of video images was possible only through the use of expensive or complex telecommunication systems, e.g., satellite, microwave dishes. Recent developments in technology and telecommunications made it quite feasible to transmit huge amounts of text, images, audio and video (multimedia) collectively over simpler networks. Electronic mail (store-and-forward) technology was utilized in less prioritized clinical applications (Perednia & Allen, 1995; Wayman, 1994). Another contributor to the recent growth in TMVC was the role played by the equipment suppliers in aggressively promoting their VC products (Perednia & Allen, 1995).

Recent technological developments led to the creation of different telemedicine innovations, such as electronic stethoscopes, odoscopes, opthalmascopes, palpation sensory transmitting gloves and others (Wayman, 1994).

In a recent review of telemedicine activities undertaken prior to 1993, Perednia and Allen (1995) found that none of the programs begun before 1986 has survived. Although data is limited, the early reviews and evaluations of those programs suggest that the equipment was reasonably effective at transmitting the information needed for most clinical uses and that users were for the most part satisfied. However, when external sources of funding were withdrawn, the programs disappeared, indicating that the single most important cause of their failure was the inability to acquire continued financial support.

THEORETICAL FRAMEWORK

It was anticipated that relying on technological innovation theories would provide an insight into factors that would influence TMVC adoption. Scarce research exists which tackles factors influencing IS/IT adoption in New Zealand (Bacon, 1992; Elliot, 1996). Most recent research (NZHIS, 1995a; 1995b; 1996; Neame, 1995) tackling New Zealand HHSs highlights hindrances at the organizational levels and at the strategic levels for information systems/technology (IS/IT) planning. This research uses the technological innovation theory as a potential framework in guiding the research procedure. In this context, the adoption of TMVC is viewed as adoption of an innovative idea.

Innovation can be defined as an idea, practice or product that is perceived as new by the potential adopters even if it had existed earlier elsewhere (Rogers, 1995). The recent emergence of telemedicine in the early nineties, due to technology advancement and to intensive technology push by suppliers (Perednia & Allen, 1995), is an innovation. Video conferencing (VC) is part of communication and automation technology, thus part of meta-technologies and information technology (IT) (Loveridge & Pitt, 1990). VC utilizes video cameras, computers or televisions and telecommunication-based technologies in transmitting audio and video images over high-speed links and networks in real-time mode.

Elliot (1996) indicates that most strategic information systems planning models have focused on the alignments of business objectives and IT planning and management of an application portfolio. A few models existed that highlighted the decision-making process for IS/IT projects. He found one model introduced by Bacon (1992) (conducted on 80 large companies in the U.S., UK, Australia and New Zealand), which highlighted 15 influencing factors on the decision to adopt IS/IT projects. These essential factors could be categorized under two main subheadings: support of business objectives and financial feasibility (e.g., discounted cash flow, net present value). Rogers' (1983; 1995) model appeared to be the most widely accepted model by researchers in identifying critical characteristics for innovation within a social system (Premkumar & Roberts, 1999; Moore & Benbasat, 1996; Thong, 1999). The social system refers to a closed and well-defined set of actors who may or may not react to each other's actions. The same researchers contend that Rogers' (1995) innovation characteristics (relative advantage, compatibility, complexity, trialability and observability) should be blended with other contexts to provide a more holistic adoption model—contexts like organizational, individual and environmental. Larsen and McGuire (1998) indicated that Rogers' (1995) model is only applicable to innovations that do not change their essential characteristics during the diffusion process. Unlike IS/IT that needs customization according to customers' needs during the introduction and diffusion phases, video conferencing technology has fixed characteristics and those do not change during the diffusion process, which justifies this research into TMVC using Rogers' (1995) model. Tornatzky and Klein (1982) examined the relationship between innovation characteristics and adoption. Such findings are still valid and endorsed by recent research in IS/IT adoption literature (Premkumar & Roberts, 1999; Thong, 1999). They highlighted Rogers' (1995) innovation characteristics and introduced the effect of the "cost" factor on innovation adoption. Although Rogers (1995) indicated that the effect of "image" could be tackled from within the relative advantage factor, Moore and Benbasat (1996) emphasized "image" as a factor on its own. The image factor was found important to the adoption of technologies in the health literature (Little & Carland, 1991). On the other hand, Rogers' (1995) compatibility characteristic is highly envisaged here, as past studies (Austin, 1992; Austin et al., 1995; Wayman, 1994) have considered the problem relating to physicians accepting information technologies (ITs) for clinical purposes.

However, it would be quite difficult within the scope of this research to explore the effects of all contexts on TMVC adoption (innovation, organizational, individual and environmental). Driven by the above argument about the applicability of Rogers' (1995) model to TMVC, it was decided to limit those to technological innovation context only. The characteristics of the technological innovation context are summarized in Table 1. As observed, the social perspective within Rogers' (1995) definitions of the different factors (Table 1) is quite apparent.

This case thus limits the study of factors affecting adoption of TMVC only to a few factors: relative advantage, complexity, compatibility, trialability, observability, cost, and image as shown by Figure 1

The research outcome is expected to add to the existing literature on adoption of complex technology such as TMVC for psychiatry purposes.

*Table 1: Innovation Characteristics**

1	Relative advantage: the degree to which using technology is perceived as being better than using its precursor of practices.
2	Complexity: the degree to which technology is perceived as being easy to use.
3	Compatibility: the degree to which using technology is perceived as being consistent with the existing values, and past experiences of the potential adopter.
4	Trialability: the degree to which technology may be experimented with on a limited basis before adoption.
5	Observability: the degree to which the results of using technology are observable to others.
6	Image: enhance one's image or status in one's social system.
7	Cost: the degree to which technology is perceived as cost effective (Tornatzky and Klein, 1982)

* Points 1, 2, 3, 4, 5, 6 are as defined by Rogers (1995).

Figure 1: Technological Innovation Impacts on Telemedicine Adoption

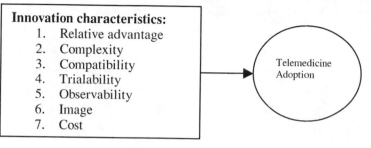

Further identification of those factors that facilitate and those that hinder adoption of TMVC would not only assist the authorities of the organization under study but also help other HHSs and policymakers.

SETTING THE STAGE

TMVC emerged (provided by one of the interviewees) in New Zealand in 1993 with a study funded by the Ministry of Communications, which was implemented by an organization called the World Communication Laboratory (no longer exists) to identify areas where New Zealand might play a role in the development of the information society. Telemedicine was one of the enabling technologies explored. One of the early initiatives emerged in 1993 within Northland HHS in transmitting radiology images between two hospitals using leased telephone lines. A consulting company called Telemedicine New Zealand Limited (no longer exists) started in 1994 to study the feasibility of telemedicine within New Zealand.

KiwiCare is a leading HHS in New Zealand. It provides a range of community and mental health services. KiwiCare maintained a simple and almost flat organizational structure with five directorates as shown by Figure 2.

The general managers of each directorate reported directly to the chief executive officer (CEO) of KiwiCare. KiwiCare Mental Health Services (KMHS) was one of the five directorates and covered mental patients at the

Figure 2: The Organizational Structure of KiwiCare in 1996

two hospitals and at the rural community centre as shown by Table 2. The rural centre is a one-hour drive away from the main hospital, and psychiatrists used to visit the rural mental health customers regularly. For providing effective mental health services to the rural area, KiwiCare has to either establish a full psychiatry centre or ensure regular visits by the psychiatrists. KiwiCare has decided on the latter course due to lack of finances to setup a full psychiatry centre. The introduction of the TMVC systems was intended to eliminate even those regular visits by psychiatrists, so as to enable them to concentrate on their clinical responsibilities.

In 1995, KMHS identified the need for a solution to the rural community mental health needs and the various options available in establishing mental health services there. The TMVC was introduced in 1996, and all the consultants and registrars have used the TMVC for general purposes of meeting and training across the two hospitals and not for clinical purposes. However, two out of 22 psychiatrists have been dedicated for the clinical use of TMVC for the rural centre. The rural centre has three nursing staff serving 100 mental health customers.

CASE DESCRIPTION

The clinical director of KMHS is a member of a regional society for psychiatrists in Australia. In 1995, the clinical director attended one of the society's conferences where the benefits of TMVC were presented. He also witnessed the usage of TMVC in mental health services and its wide usage in Australia. Upon his return he shared the idea of introducing TMVC in his hospital with his general manager.

Table 2: Organizational Information About KiwiCare

Description	KiwiCare
Number of hospitals	Two hospitals: 400 beds (Headquarter) and 60 beds
Number of rural centres/hospitals served	1 rural community centre
Number of employees	Around 3000
TMVC initiation	Late 1995
TMVC adoption	Mid. 1996
Number of TMVC systems installed	3 group video conferencing systems at the two hospitals and at the rural community centre
TMVC clinical application	Telepsychiatry between the headquarter and the rural community centre
Bandwidth (ISDN: integrated systems digital network)	128 then later on to 384 kbps

The general manager of KMHS, who is 44 years old and has been working in the health industry since 1975, has shown keen interest in the project. Various benefits that would accrue by introducing TMVC were identified: the system allowed instant and continuous access to remote rural patients; a solution to their recruitment problem in finding psychiatrists accepting working in the rural community centre; saved on psychiatrist's time lost in travelling to see more patients and reduced their workload; could be used for managerial meetings and for reviewing medical journals, either with rural community centre or with the other hospital; and could be used for medical training (continuing education). The general manager indicated that KMHS desired to be a leader in the use of TMVC in mental health services in New Zealand.

In order to provide proper treatment to psychiatry customers, seeing them in person (one-on-one) was essential. Visual images of the patient facial reflections and their sounds were essential and could not be compromised. The use of other technologies like audio conferencing alone or video images or photos were inadequate. However, basic follow-up techniques utilized by KMHS psychiatrists had relied heavily on telephone conversations (i.e., whether patients were taking their drugs regularly). TMVC was seen as the only acceptable solution for psychiatry purposes in seeing patients and for follow-up.

Seventy percent of KMHS psychiatrist activities were based on home visits (one-on-one) to their mental health consumers, where the actual examination took place. Patients can be seen in their real environment: where they live, the people surrounding them, etc.

KMHS relied on psychiatrists to move to the TMVC room to operate the system and to establishing the connection with the rural centre. There were no dedicated staffs for the TMVC system installed.

As per the discussions the general manager saw an opportunity by adopting TMVC and presented this TMVC project to the top management, which highlighted the financial gains along with other advantages. KiwiCare like any other HHS is under tight financial constraints. The project plan anticipated to save NZ$100,000 annually on psychiatrist time and on travelling expenses. It was a cheaper option than establishing a psychiatry centre in the rural area. The plan envisaged implementation of the project on a lease-to-buy basis rather than requesting the whole amount of money in advance from the top management. He even arranged the funds for the project through internal funding. In addition to clinical advantages, it could be used for various managerial and training purposes.

Since it was a novel project, there were concerns regarding the availability of a technical person who could understand and maintain the ongoing operation of the system. The supplier promoted their VC product aggressively to KiwiCare. They demonstrated the basic features of the system to the clinical director and to two psychiatrists by dialing into the regional head office of the supplier in Australia. The general perception gained was the system was not complex and easy to operate with minimal training. There were not many technically difficult operations, but only some minor techniques such as controlling the camera (scanning, zooming) and use of a remote control device. KMHS in this regard made sure to take a guarantee from the VC supplier for the ongoing support and systems availability.

Finally, after approval was obtained from top management, the TMVC project was adopted in 1996. Three VC systems were installed in the two hospitals and in the rural community centre with the ISDN 128 kbps bandwidth. Each system was placed in a separate room equipped with electrical and data points and a good lighting system. After conducting the training on the system by the supplier, the system was put in use.

After adopting the system various difficulties were highlighted by clinicians. The clarity of the images and the sound weren't acceptable to the psychiatrists in seeing their patients. This has hindered the frequent usage of the system. Consequently the bandwidth was upgraded from 128 kbps to 384 kbps, where the clarity of the images had been enhanced, although the sound quality did not improve to the acceptable standards. Sometimes the TMVC encountered lots of blackouts due to the unreliable ISDN connection and related peripherals such as network-bridge, which crashed under overloading. Likewise the bandwidth and the VC suppliers encountered difficulties in finding solutions to such technical problems. The lack of a dedicated help-desk facility to attend to psychiatrists' urgent technical calls created lots of frustrations even to the psychiatrist-registrar during the period he was managing the system.

Further, psychiatrists found it burdensome to leave their office and conduct a session. They had to learn manoeuvring the camera across the room to interview the mental health customers and their families accompanying them. Getting mental health customers to stare at the camera rather than at the television screen was also highlighted. Fear of the technology (i.e., afraid to touch the VC and damage expensive equipment) and the dislike of seeing one self on the screen were witnessed among the computer-illiterate clinical staff and even among the elderly doctors. In one instance one of the nurses in the rural community centre almost fainted when she saw herself on the screen.

Psychiatrists developed a strong opinion about TMVC (one-to-one) interfaces as being not acceptable in the psychiatry area. Issues concerning patient's reactions, hand movement, facial reflections and above all the home visits and the one-on-one basis were quite prevalent among psychiatrists. If a psychiatrist was present at the rural centre, there was no advantage in having TMVC consultation, simply because the psychiatrist was a specialist. Seeing new patients was always emphasized to be one-on-one and in person, and locating junior psychiatrists in rural areas supported by TMVC uplinks was not recommended and was quite risky clinically and legally. The situation was further worsened with the technological complexities and failures, which created lots of frustrations among the TMVC users in general and the lack of a dedicated help-disk facility to attend to urgent faults.

As the clinical director indicated, although telepsychiatry is successful in Australia, the Australians are ahead of New Zealand in TMVC by many years and developed the hands-on skills in going along with the technology.

Some of the psychiatrists were enthusiastic about the technology. This was supported by the fact that there were some psychiatrists who were consistently using the technology in seeing their patients specifically the ones in the rural area. This was driven in part by the relative advantage of the technology indicated earlier and by the fact they had showed a personal interest in technology and in developing and mastering the man-machine interaction. When those psychiatrists left KMHS, the hospital had to revert back to their earlier practices in seeing rural mental health customers in person by other psychiatrists.

One year after adopting the TMVC system, KMHS was able to attract a donation from the local bandwidth provider to sponsor a psychiatrist-registrar for one year to manage and to empower the TMVC initiative within KMHS. That is to implement and execute a protocol that would coordinate TMVC sessions and schedule, and at the same time, to explore other opportunities where TMVC could be further utilized. For instance TMVC was also used in conducting interviews for recruitment purposes, and for judicial reviews, where judges had used VC technology to review the status of mental health customers and for various legal proceedings. At these instances, the judicial encounters with patients were recorded on videocassettes and kept securely as legal evidence. Other than that, the regular TMVC encounters were not recorded due to privacy and legal barriers. Also on some occasions, the system was rented to rural businesses to conduct their VC sessions with overseas countries. KMHS was not able to renew the donation, and the registrar had to go back to his earlier duties as a psychiatrist. The registrar indicated that he spent most of his time in responding to technical queries from psychiatrists who were using the system and in coordinating the technical visits for the TMVC supplier. The registrar concluded that TMVC should be fast, intuitive, robust, stable and trustworthy (FIRST) before it could be relied on for clinical purposes.

Patients' perceptions were not considered before the adoption decision, but after adoption results showed that rural patients were curious and comfortable with the technology and saw it as easier than travelling to the main hospital.

CURRENT CHALLENGES/PROBLEMS FACING KMHS

According to the case description and after portraying the impact of the innovation characteristics on TMVC before and after its adoption by KMHS, Table 3 summarizes the perceptions made about TMVC prior to its adoption and the use of the system after adopting it. This was essential in highlighting the importance of the innovation characteristics on the adoption decision for TMVC.

In lieu of Table 3 factors like relative advantage, cost effectiveness, observability, and image were the main contributors to TMVC adoption. The first two factors were the main factors behind TMVC adoption, and the rest acted as facilitators and accelerators. This substantiates Bacon's (1992) findings, which indicates that organizations adopt IS/IT projects based on their support for explicit business objectives and on their cost effectiveness. But as observed by the research literature and by findings from the case that this is not sufficient to guarantee successful adoption of TMVC and its subsequent utilization afterwards as the system was used minimally.

KMHS has undertaken a cost-benefit analysis to justify the investment made on TMVC. Other factors like complexity, compatibility and trialability (which could capture issues arising from complexity and compatibility) weren't tackled rigorously. These conclusions were backed by findings from the current utilization of TMVC within KMHS. This may have been justified in part due to the fact that the technology was newly introduced in New Zealand and there is not much knowledge about it among suppliers and potential adopters. That's why early adopters of the technology face higher risks and encounter higher expenses in gaining the technology and the know-how. Other hospitals interested in TMVC will benefit from the reduced prices and the advancement in the technology and the experiences of early adopters. This could be gained from the adopter directly or from the vendor who developed the local expertise and already has an edge over other suppliers or from research studies like the current one.

Table 3: Impact of Technological Innovation Factors on TMVC Before and After Adopting it by KMHS

I	Innovation Characteristics	Before Adoption	After Adoption
1	Relative advantage	Highly identified as outlined earlier.	Not fully utilised and affected by other factors such as complexity and compatibility factors.
2	Cost	Justifiable and the business plan was a successful.	KiwiCare could not afford to maintain a dedicated registrar to empower the ongoing operations of the TMVC project.
3	Complexity	Easy to use with basic training	Complex technology, not easy to use and to administer, technical knowledge was needed. Technology was not reliable, the bandwidth and network bridges used to trip and crash.
4	Compatibility	Not concerning and therefore was not explored thoroughly. The earlier perceptions were with the system.	Not compatible with earlier practices and values. The one-on-one encounters seemed essential and could not be compromised.
5	Trialability	No concerns/not explored thoroughly	If tried before adoption it would have detected complexities and incompatibilities issues pertaining to TMVC.
6	Observability	Contributed to the unanimous agreement on TMVC usefulness. TMVC is successfully used in Australia.	Although it was adopted but if KMHS reviewed relevant literature about TMVC in the psychiatry area they would have observed the true (or lack of) usage of the technology for clinical purposes.
7	Image	It will further enhance KMHS's image. It will also endorse its leadership in mental health services as well.	KMHS was approached by other HHSs in New Zealand to benefit from their experience with the TMVC project.

The inability of KMHS to dedicate a registrar for empowering the TMVC activities resembled a challenge in justifying the investment made (cost/benefit) on TMVC. Prednia and Alen (1995) highlighted the failure in justifying the cost-benefit analysis by hospitals. On the other hand, adopting and implementing a protocol in coordinating TMVC sessions and schedules is of paramount importance to the success of the TMVC project as it allows for better cost control and efficient time-allocation. Therefore, including it as part of the feasibility study is essential.

Rogers' (1995) compatibility characteristic has revealed essential social and cultural implications within the psychiatry practice at KMHS. TMVC seemed to penetrate core clinical paradigms within the psychiatry practice. Issues like one-on-one, in person, facial and body reflections, hand movements and patient real environment setting were all practiced in person by psychiatrists in order to achieve a better rapport with their patients to achieve effective treatment. Using VC seemed to create lots of rejection to the technology among psychiatrists. This has been further aggravated by the failure of the technology and the bandwidth in supplying reliable and clear images and sounds of the patients (complexity characteristic).

The possibility of high risk involved in misdiagnosing and in mistreating patients in using TMVC that could lead to tragic consequences and legal liabilities has further limited its usability among psychiatrists and KMHS. This in fact eliminated one of the essential utilities of TMVC as a clinical tool and its importance within a rural setting. This highlights a disadvantage and incompatibility issues concerning TMVC adoption among psychiatrists and the importance of the preparatory stage to the adoption and use of TMVC among users. Wayman (1994) has pointed to this perspective and to the importance of teaching doctors first about TMVC before introducing it. This gradual process is highly envisaged here.

We have seen that Rogers' (1995) innovation characteristics supported by other technological innovation characteristics from the literature has helped in gaining a richer picture about factors influencing TMVC adoption within KiwiCare and in identifying the most influential ones, thus, validating the applicability of technological innovation theories in providing an understanding about TMVC (VC) adoption within KiwiCare. It was observed that some relationship exists between the depicted technological factors. For instance if the cost-benefit analysis found TMVC an expensive option, KMHS would have considered this a disadvantage and rejected the project. The complexity of TMVC has frustrated psychiatrists and further aggregated their rejection (incompatibility) to the technology. Therefore, this research emphasizes the importance of exploring the type of interrelationships that may exist between the different innovation factors. This is essential and would influence the adoption decision.

Accordingly, Figure 3 suggests a framework that would guide KMHS and other HHSs in their potential uptake of TMVC or any other complex technology.

This research further highlights the need to explore the effects of other contextual factors (organizational, individual, environmental) on TMVC adoption. Telepsychiatry is being used successfully in countries like the U.S. (Perednia & Allen, 1995; Wayman, 1994) and Australia, as indicated by one of the interviewees. This highlights the importance of considering factors other

Figure 3: An Adoption Framework for TMVC

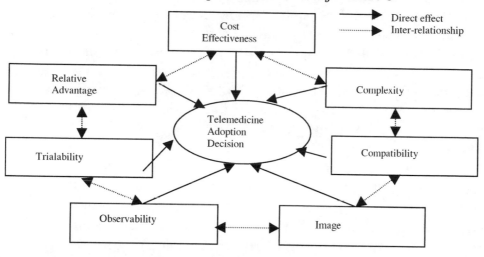

than the technological ones. Factors like the organization and the environment surround the adoption of the technology. For instance the general manager's leadership in facilitating TMVC adoption within KMHS was quite apparent. Expanding on this issue is necessary and will highlight essential features of entrepreneurs and product champions within organizations in facilitating not only the adoption of TMVC but the post adoption as well. This envisages a better role to be played by KMHS, other HHSs and health policy makers in assessing the real potential of TMVC alongside essential characteristics highlighted by the technological innovation theories.

CONCLUSION

In line with the technological innovation theories, specifically Rogers' (1995) theory has proven robust in detecting factors influencing the adoption (or rejection) of technologies like the VC in the case of the KMHS TMVC initiative. Thus, allowing for the social perspective to emerge alongside the different factors. Cost-benefit analysis is a classical managerial practice in assessing the potential of adopting projects in general. It is essential in the first place, but a more rigorous approach should be adhered to in dealing with new and expensive technologies like the VC and in predicting its operational costs afterwards. The context within which the technology is introduced to its potential adopters should consider other important factors and contexts in order to look at the technology from different angles. Issues like the complexity of the technology and whether we have the local experts and expertise to overcome adoption and implementation barriers should be considered. Another highly emphasized issue is whether the current working environment to accept new changes imposed by the new technology should be studied in advance. The last two issues could be easily detected in advance if we are allowed to trial with the system for a period before adopting (or rejecting) it. This is not a strange request and it is offered by the technology vendor industry. This further emphasizes the importance of the depicted framework shown in Figure 3, which could be supplemented by a further study to explore the effect of interrelationships among the innovation factors on the adoption decision. This could be the scope of a large empirical research.

These are the main themes explored within this study and it is up to other researchers to explore the effects of other contexts, factors and inter-relationships. At the micro-level, issues like training, end-user involvement, empowerment and top management support, leadership and motivation are highly envisaged here. At the macro-level, however, Perednia and Allen (1995) provided various implications, although not all of them would necessarily apply to New Zealand HHSs, e.g., licensure. Above all considering the option of integrating TMVC completely within core clinical activities will further contribute to its success, rather than designing it at the outside and then invite users afterwards to operationalize the system.

This should provide good grounds in understanding the adoption procedure for TMVC in the case of KMHS. It is up to KMHS and other HHSs and policy makers in understanding the effects of the various factors on the adoption of complex technologies like TMVC.

FURTHER READING

New Zealand Sites

New Zealand Hospitals: http://www.nzhealth.co.nz/
New Zealand Health Funding Authority: http://www.hfa.govt.nz/hfahomepage.htm
New Zealand Ministry of Health

General Sites

Telemedicine: http://tie.telemed.org/journals/
Databases: http://www.brint.com/, http://www.nua.com/

Video Conferencing Vendors

VTEL: http://www.vtel.com
Polycom: http://www.polycom.com/

Telemedicine Journals

Journal of Telemedicine and Telecare: http://www.roysocmed.ac.uk/pub/
jtt_ed.htm
Telemedicine Journal and E-Health:
http://www.liebertpub.com/TMJ/Manuscripts/default1.asp
New Ethicals Journal: *New Zealand's Journal of Patient Management.*

Telemedicine Magazines

Telemedicine Today: http://www.telemedtoday.com/
Telehealth Magazine: http://www.telehealthmag.com

REFERENCES

Anderson, J. (1997). Clearing the way for physicians: Use of clinical information systems. *Communication of the ACM*, 40(8), 83-90.

Austin, C. (1992). *Information Systems for Health Services Administration.* Michigan: AUPHA Press/Health Administration Press.

Austin, C., Trimm, J. and Sobczak. P. (1995). Information systems and strategic management. *Healthcare Management Review*, 20(3), 26-33.

Bacon, C. (1992). The use of decision criteria in selecting information systems/technology investments. *MIS Quarterly,* September, 369-386.

Elliot, S. (1996). Adoption and implementation of IT: An evaluation of the applicability of western strategic models to Chinese firms. In Kautz, K. and Pries-Heje, J. (Eds.), *Diffusion and Adoption of Information Technology*, 15-31. London: Chapman & Hall.

Grigsby, B. and Allen, A. (1997). Fourth annual telemedicine program review. *Telemedicine Today*, 30-42.

Larsen, T. and McGuire, E. (Eds.) (1998). *Information Systems Innovation and Diffusion: Issues and Directions*. Hershey, London: Idea Group Publishing.

Little, D. and Carland, J. (1991). Bedside nursing information system: A competitive advantage. *Business Forum*, Winter, 44-46.

Loveridge, R. and Pitt, M. (Eds.). (1990). *The Strategic Management of Technological Innovation*, London: John Wiley & Sons.

Moore, G. and Benbasat, I. (1996). Integrating diffusion of innovations and theory of reasoned action models to predict utilisation of information technology by end-users. In Kautz, K. and Pries-Heje, J. (Eds.), *Diffusion and Adoption of Information Technology*, 132-146. London: Chapman & Hall.

Neame, R. (1995). *Issues in Developing and Implementing a Health Information System*. Wellington, NZ: Ministry of Health.

New Zealand Health Information Service (NZHIS). (1995a). *Health Information Strategy for the Year 2000: Stocktake of Current Position and Future Plans*. Wellington, NZ: Ministry of Health.

New Zealand Health Information Service (NZHIS). (1995b). *Health Information Strategy for the Year 2000: Gaps, Overlaps, and Issues Report*. Wellington, NZ: Ministry of Health.

New Zealand Health Information Service (NZHIS). (1996). *Health Information Strategy for the Year 2000*. Wellington, NZ: Ministry of Health.

Office of Technology Assessment U.S Congress (OTA). (1995). *Bringing Health Care On Line: The Role of Information Technologies*, OTA-ITC-624. Washington, DC: U.S. Government Printing Office.

Perednia, D. and Allen, A. (1995). TMVC technology and clinical applications. *The Journal of the American Medical Association (JAMA)*, , 273(6), 483-488.

Premkumar, G. and Roberts, M. (1999). Adoption of new information technologies in rural small businesses. *The International Journal of Management Science (OMEGA)*, 27, 467-484.

Rogers, E. (1983/1995). *Diffusion of Innovation*. New York: The Free Press.

Thong, J. (1999). An integrated model of information systems adoption in small business. *Journal of Management Information Systems*, 15(4), 187-214.

Tornatzky, L. and Klein, K. (1982). Innovation characteristics and innovation adoption implementation: A meta-analysis of findings. *IEEE Transactions on Engineering Management*, 29(11), 28-45.

Wayman, G. (1994). The maturing of TMVC technology Part I. *Health Systems Review*, 27(5), 57-62.

BIOGRAPHICAL SKETCH

Nabeel Al Qirim has been working as a consultant in the IT industry since 1989 with companies like IBM, Compaq, Siemens Nixdorf and Data General. He managed different IT projects and turnkey solutions. He has taught e-commerce at the UNITEC Institute of Technology and the Auckland University of Technology, New Zealand. His research interest is in the adoption and diffusion of complex technologies, telemedicine, e-commerce and small business. He has a bachelor's a degree in Electical Engineering (BEE), Information Systems (Hons.) and an MBA.

The Role of Virtual Organizations in Post-Graduate Education in Egypt: The Case of the Regional IT Institute

Sherif Kamel
American University in Cairo, Egypt

EXECUTIVE SUMMARY

Learning, education and training using traditional class methods and/or emerging online techniques are all leading to improved ways to investing in people, leveraging their capacities and reaching out to remote masses while cutting down on cost, time and efforts. Thus, the role of virtual organizations and virtual teams is rapidly spreading worldwide in the related aspects to learning and human resources development. This has led to the establishment of a large number of regional and global learning consortiums and networks aiming to provide quality knowledge and information dissemination vehicles to an ever-growing community of seekers that is online, active and eager to increasingly learn more. However, most of the publications address the theoretical foundations of virtual organizations; as for the actual practices, they are not extensively reported in the literature. This case addresses the experience of Egypt's Regional IT Institute in the field of education and training. Today, the learning process is becoming a vital factor in business and socioeconomic growth where the role of information and communication technology is having a growing and an innovative impact. The Regional IT Institute was established in 1992 targeting the formulation of partnerships and strategic alliances to jointly deliver degree (academic) and non-degree (executive) programs for the local community capitalizing on the enabled processes and techniques of virtual organizations. The case provides many lessons to be replicated that demonstrate the opportunity to expand in exchanging the wealth of knowledge across societies using a hybrid of forms for virtual organizations and virtual teams.

INTRODUCTION

New online learning and education techniques are leading to improved ways to investing in larger numbers of people and to leveraging their capacities with fewer resources. Respectively, virtual organizations are gaining grounds in the educational sector. However, the essence is to provide mechanisms for knowledge acquisition and dissemination to a growing global society of learners and trainees. This is an important vehicle to close the global digital divide and a mechanism to reach out to the remote areas worldwide. Egypt has had a number of experiences in that direction in an attempt to leverage its literacy rate while making optimum use of its limited resources. This chapter covers the case of the Regional IT Institute, an institute located in Cairo and specialized in providing quality education in the field of information and communication technology and management. However, it is important first to explore the environment in which the institute has been operating for almost a decade.

Egypt is the cradle of an ancient civilization dating back to 3000 BC. With a population of about 68 million, out of which 19 million are in its workforce, Egypt is the most populous country within the region. About 24% of its population is in the education stages. It has the second largest economy in the Middle East and has successfully implemented its economic reform program that has enabled its current economic growth rate to stand at 6.2% annually with an inflation rate of 2.1%. Egypt has four basic sources of foreign exchange earnings equally divided between tourism, oil, Suez Canal earnings and remittances of Egyptians working abroad (Kamel, 1999). Cairo, the capital of Egypt, is a large metropolis where buildings of French and English architecture stand next to modern skyscrapers. Cairo's 18 million inhabitants constitute nearly 27% of the total population. Egypt, like many other developing countries, is trying to expand its industrial base and modernize itself technologically. Agriculture accounts for 15% of the gross domestic product and industry accounts for 40% with a large service sector mainly built around tourism and transportation. Its major exports are human resource capacities, petroleum products, cotton and leather products, while its major imports include food, machinery and vehicles. This is coupled with a literacy rate of 61%, a computer literacy rate of 8%, and a per capita income of U.S.$1,465 (Kamel, 1999).

Computing was introduced for the first time in Egypt in the 1960s but it was not until 1985 that information technology was put on the national agenda and became one of the building blocks of the overall business and socioeconomic development process. In 1999, a new ministry was established and added to the cabinet portfolio to help accelerate the efforts for building an IT-literate society capable of competing on a global scale and keeping pace with the new trends of the digital economy: the ministry of communication and information technology. The new ministry's agenda is to build the nation's information and communications infrastructure and on the top of the agenda is investing in people and formulating a network for knowledge to help bridge the gap between the haves and the have nots in Egypt and also between Egypt and other developed nations to keep pace with the developments taking place elsewhere in the world. However, with an economy in transition and the availability of all the classical problems of developing nations, including scare infrastructure resources and limited funds, it was inevitable to capitalize on other experiences and start from where others have ended to be able to realize an impact in the shortest time possible. Therefore, a vertical development strategy was set based on using communication and information technology as a backbone for development and respectively capitalizing on the various techniques of the digital economy.

The continuous innovation in information and communication technology led to the development of virtual organizations with different forms and structures. Virtual organizations are defined as networks of institutions that, using cutting-edge technology, unite to provide a value-added competitive advantage (Strausak, 1998). The virtual integration of its respective capacities, irrespective of time, effort and distance barriers, enables the realization of its common objective: becoming more competitive in the local as well as global marketplace. The ultimate goal of a virtual organization is to achieve market differentiation and better performance through the use of information technology (Appel and Behr, 1997). Moreover, virtual organizations, if well implemented, can help realize more with less via consolidation and rationalization of resources allocation and use. The world is currently living in an emerging knowledge-based global economy where the continuous innovations in information and communication technology have forged stronger links between individuals, organizations and nations (Ungson and Trudel, 1998). This is leading to the creation of growing opportunities for collaboration in various fields and helping in the adaptation and management of technology to serve various purposes and objectives (Palmer, 1998). In the past, organizations worldwide were competing in trade and raw materials; however, during the next millennium the focus will be on intellectual property rights and globally oriented products and services. The 21st century will address more knowledge and information-intensive issues where the critical element is people "humanware" (Kamel, 1998a). People and their invaluable contribution to the society will be key for development and growth and its intellectual capital that will prevail in the global digital economy of the 21st century.

With growing competition in information societies, developing standards and reducing costs, virtual organizations represent the opportunity in the 21st century to reallocate resources and to reposition organizational status in the global cyber marketplace aiming to supply goods and services by means of its staff, equipment and information systems (Sieber, 1997). The changing economical, business and technological trends and the need to rationalize the use of resources has led to the evolution of the concept of virtualness. The virtual organization, as a concept, is based on strategic considerations such as optimizing the business size, identifying the market area and ascertaining core competencies (Sieber, 1997). Training and education are vital societal requirements that are reflected by the growing need to train and educate more people continuously and on diversified subjects, meaning an expansion that has both vertical and horizontal dimensions. However, with a limited infrastructure and scarce financial resources, such an objective becomes difficult to achieve (Kamel, 1999). This factor is magnified in developing economies where limited resources are usually allocated to sectors such as food and health as a priority to education. However, things are changing as the world is moving from focusing on products and manufacturing to knowledge and investing in people. Countries of the world will focus on developing new methods to absorb the growing learning needs of the society, and on managing knowledge, otherwise their business and socioeconomic development plans will be difficult to realize and developing countries will risk losing more grounds to the developed world (Kamel, 1998b).

Flexibility is an important element of organizational virtualness (Kocian, 1997), and in the education and training fields, diversity of domains and scheduling of topics provide a wider choice for the beneficiary enabling a larger pool of specialties and professions. The use of information and communication technology has been a key factor in the success of the concept of virtual organizations. The availability of the World Wide Web and the Internet facilitated the exchange of information and knowledge between different partners and provided new mechanisms for the provision of value-added outcomes using information and communication technology (Gristock, 1997). Moreover, the new competitive differential depends on the creation and application of technology as a form of knowledge (Ungson and Trudel, 1998). It also depends on how successful this knowledge is being managed to be able to be accessed by the right people and at the right time from anywhere around the globe.

EGYPT AND THE KNOWLEDGE ECONOMY

Education and training are key success factors in societal growth. Moreover, the trends of globalization, competition and the increasing role of information and communication technology are empowering the waves of global change in our societies in the 21st century. These trends require the emergence of highly skilled and qualified human resources. Therefore, it is urgent to invest in people and to build new generations capable of meeting market and industry challenges. The 21st century will create a knowledge-based society where the fundamental sources of wealth will be knowledge and communication rather than natural resources and physical labor, and it will be up to different societies to identify the best formula that meets its requirements and realizes an optimum return in terms of information dissemination and knowledge management (Ungson and Trudel, 1998). Therefore, since 1985 Egypt has invested heavily in human resource development through two dimensions: education and training. Thus, over 1,000 training centers were established to address management and IT issues linked to the needs of the market and covering various sectors in the economy across Egypt's 27 provinces. These centers had a remarkable impact on the development of human capacities, skills and knowledge. The achievements to date include a large number of programs that contributed in leveraging the skills and knowledge for many fresh graduates as well as employees across different organizational levels.

With 14 million students in the education sector in Egypt, the challenge is to develop modalities to be able to educate and train more people while optimizing the allocation and use of available resources. Therefore, building a virtual learning model represents one of the possible vehicles that could realize such an objective. Therefore, the aim of the model is the development of strategic alliances with learning institutions around the world to deliver degree and non-degree programs for the market in Egypt using state-of-the-art information and communication technology. The model builds on three main directions: home computing which serves in the knowledge dissemination and documentation; electronic mail which provides an open channel between instructors, students and institutions; and finally the World Wide Web, which is to date the best information-retrieval vehicle worldwide that is reaching users everywhere and anytime in a user-friendly format. One of these centers is the Regional IT Institute located in Cairo and working as a base for a satellite of programs in cooperation with a multiplicity of world leading institutions. The Regional IT Institute was established in 1992 to support in transforming the society using the latest technologies and methods in education and training and having as a motto building through learning. The institute is established for the design and delivery of degree and non-degree professional programs to leverage the capacities of human resources in Egypt. The programs are jointly delivered with the Institute's strategic alliances represented by international institutions that disseminate knowledge through the use of hybrid methods, including class sessions and distance-learning techniques. The case is built around trust between the involved parties (Brigham and Corbett, 1996). At the same time, with neither a hierarchy in place nor a leading role played by any of the involved parties (Appel and Behr, 1997). The model helps setting recommendations for similar initiatives that address one of the global growing needs: human resource development.

CASE STUDY: REGIONAL IT INSTITUTE (RITI)

The Regional IT Institute (www.riti.org) was established as a subsidiary of the Regional Information Technology and Software Engineering Center (RITSEC). It is a not-for-profit organization supported financially by its various services and programs that target the education and professional development discipline as an important vehicle for business and socioeconomic development. The Regional IT Institute has become one of the leading institutions in Egypt in degree and non-degree education in the fields of business, management development, computing, and communication and information technology. Today, the Institute extends its services through a global outreach program to countries in Europe, Africa and Asia through the development of joint alliances with regional and international institutions. The Institute pioneered in Egypt in introducing a new mode of operation that is based on virtual teams and that led to date to the enrollment of 806 students, 382 of whom have already graduated with a master's degree, in addition to training more than 12,000 participants from over 1,100 organizations in approximately 90 countries in African, Europe and Asia.

The mission of the Institute is to contribute to business and socioeconomic growth in Egypt through investing in people in tailor-made executive degree and non-degree programs in the areas of business, management development, and information and communication technology.

This is meant to help close the technology gap between Egypt and the world by investing in people and keeping pace with the massive and continuous developments taking place worldwide by linking people to information and communication technology. The objectives of the Institute include: 1) investing in people as the oil of the 21st century, the societal backbone of the future and the nation's most precious resource—therefore, developed human resources will leverage IT and management capacities which will be reflected in organizational performances; 2) building world-class professionals by preparing a new IT-oriented generation that can make a difference in Egypt's growing developing economy; 3) establishing other regional IT centers of excellence by replicating the experience of the Institute to facilitate the diffusion and management of information and communication technology knowledge. Formulating market and industry-driven programs where market analyses show the lack of programs that address the actual needs of industry in IT and management domains. The synergy between the industry and the business communities and the education sector has always proved important to be able to address the needs of the marketplace, identify the required capacities, formulate the appropriate training and learning contents and processes and finally, guarantee, to an extent, employment opportunities.

The scope of services includes a portfolio of over 350 ICT and management programs divided into circles of knowledge and changing quarterly according to market needs. These circles are: business and management development, information and communication technology, and computer science and software engineering. This figure includes executive degree and non-degree programs, as well as seminars, workshops and conferences around key issues and topics important to the industry, the business community and the society at large. These services are offered to all organizational levels from business leaders and decision makers to executives and senior managers, as well as end-users including professionals and trainers from the private sector as well as government and public-sector organizations.

The Institute resides in an ornate villa, dating back to the 19th century. Stepping inside the premises, the training resources are state of the art, portraying the vital role the Institute plays in the development of the ICT human infrastructure in Egypt. Such infrastructure is continuously updated to match the ongoing developments of the information and communication technology infrastructure. It also helps in the provision of a platform that enables the proper dissemination of knowledge. The premises, totaling in area to 1,500 m², encompassing four lecture rooms and four computer labs, is equipped with advanced computing facilities with both PC and UNIX-based platforms, including 110 personal computers with their networking and printing peripherals. All computers have direct and full Internet accessibility. Most of the ICT infrastructure is heavily used in communicating with partnering institutions worldwide as well as for internal transactions and operations. The reason for that is to render the service of the Institute more efficient and cost effective as well as to save on time and efforts. The ICT infrastructure has been a deciding factor in the success of the Institute because being connected to its partners is the enabling factor for the Institute to realize organizational virtualness and in information acquisition and knowledge dissemination (Davidow and Malone, 1993; Byrne, 1993; Goldman, Nagel and Preiss, 1995). In that respect, the Institute utilizes a hybrid of technologies that encompass a mix of traditional methods for knowledge delivery coupled with a number of online tools and techniques that capitalize on emerging information and communication technology. The Institute uses mature technology such as facsimile and electronic mail as well as innovative technology such as the World Wide Web to link core competencies, serve its market and reduce the cycle time in its programs completion. Lecture rooms are fully equipped with training requirements, including overhead and slide projectors, data show, electronic white boards, televisions and VCRs. Finally, the Institute has a specialized library with books, references, videotapes and CD-ROMS in subjects related to information technology and management, as well as online subscriptions to a number of online regional and international library networks to provide local students with firsthand access to a wealth of resources that could help in their learning process.

A Virtual Learning Model

One of the success stories of the Regional IT Institute has been the joint delivery of graduate degree programs in collaboration with universities in Europe and the United States. The objectives of the programs, launched in 1994, was to provide local students with the opportunity to study for a master's degree in business administration or computer science from the partnering universities while in Egypt. The virtual learning model revolves around the selection of partners based on synergetic completion of core competencies (Kocian, 1997). Moreover, it tries through these partnerships to possess the best-in-world competencies to build a specific education service in the short term (Greiner and Metes, 1996).

The identification and selection of partners is based on the identification of universities that can cater for the educational needs of the local market in Egypt. The virtual learning model is based on the collaboration, with partnering institutions acting as small one-person firms bringing together their efforts and resources to serve the activities they jointly deliver. These institutions are drawn from different parts of the world, irrespective of their geographic location trying to globalize its virtual operation through such partnerships (Wolff, 1995; Coates, 1994). Currently, the Regional IT Institute, at the education (degree) level, jointly operates a master's degree program in business administration with Maastricht School of Management (Netherlands) launched in 1994, a master's degree program in computer science with the University of Louisville (USA) launched in 1996 and a master's degree program in business information technology with Middlesex University (UK) launched in 1998. These programs—although it all started with a single specialization area, and now based on market needs—address a lineup of specializations that include: information technology management, banking and finance, marketing management, globalization, telecommunications and Internet applications, business information technology and e-commerce.

The case of the Regional IT Institute realizes the new working world order where corner offices, paper memos and personal secretaries are out and laptops and teleconferences and periodical meetings are in (Gray, 1995). The cooperation between the Regional IT Institute and its partners could best be described as virtual arrangements with a vital role played by their joint virtual team which is one of the success factors in this operation being one of the core components of virtual organizations (Knoll and Javernpaa, 1998). The formula was new to the market in Egypt where each of the three institutions does not have an office in Egypt, however due to a virtual agreement, each institution in collaboration with the Regional IT Institute synergizes its efforts to concurrently deliver its collaborative degree in Egypt as well as in its home country, while sharing its tangible and intangible resources and capacities. The basic driving factor behind the collaboration between those teams separated by thousands of miles was the development of a dynamic system that sets the responsibilities for each institution and structures the relation between both of them (Rockart and Short, 1991). It is important to note that the collaborative effort between the Regional IT Institute and any of the three partners was totally independent from other degree programs delivered in terms of strategy formulation, management and operations as shown in Figure 1.

Although many of the activities are fully performed in Cairo or in any of the other cities (Maastricht, Louisville or London), some of the operations were centralized such as recruitment and grading. Therefore, the partnership was formulated of the hybrid type with both electronic and nonelectronic "face-to-face" collaboration. The Cairo-Partner team had to resolve its ambiguity, reduce uncertainty, establish its roles and identities and take collective actions (Nohria and Eccles, 1992). This element was vital with over 75% of the instructors not located in Egypt but rather in Germany, the United

Figure 1: Synergy Between Regional IT Institute and Partners

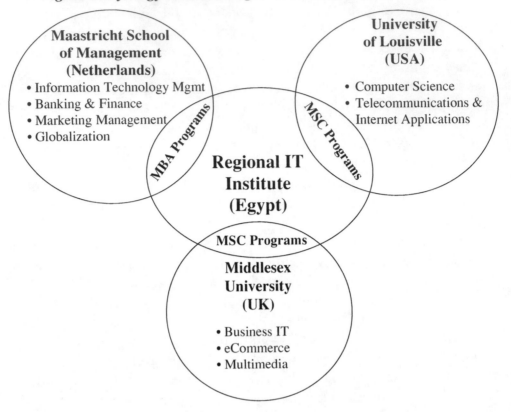

States, Canada, France, UK, India, Malta, The Netherlands and other nations. In many ways, the collaborative process is three dimensional between the instructor in country (A), the local program coordinator in Cairo and the resident program coordinator in Maastricht, Louisville or London. The institute has allocated a local program coordinator who is responsible for all transactions with a specific partner and acts as the Institute's representative in the virtual team that also includes members from the collaborating university. Virtual teaming enabled people to communicate with each other on a daily basis through computer networks, faxes and telephones, allowing a workforce to be created without relocating and irrespective of their time and distance differences. The staff in different locations was interconnected through information and communication technology using common tools to navigate the network, exchange information, students' records, grades, assignments, and sharing advice and decisions. The virtual team formula enabled a flexible and continuously evolving fit between skills, resources and growing and changing needs. Figure 2 demonstrates the relation between the Regional IT Institute and any of its collaborating partners. It is important to note that in the case of the education (degree) programs, all collaborating institutions with the Regional IT Institute are global partners. The figure shows the use of unconventional educational methods such as the World Wide Web, electronic mail and online conferencing extensively as complementary mechanisms and tools to the conventional and more traditional methods. The figure identifies the tools that were used in each method and the components that were included in each of them respectively. It is important also to note that the use of such nontraditional methods was faced with some difficulties in the beginning until the staff of both partnering organizations were adapted to the new techniques.

The learning formula is made simple to satisfy the needs of all parties by clearly setting the duties and responsibilities of each institution. Therefore, while the university takes full responsibility and accountability of all academic-related issues, the Institute focuses on the marketing, management and administration of the program. In other words, while the Institute undertakes all operations of the program, the university handles all academic elements. Respectively, the duties and responsibilities of each partner could be described as follows. On the one hand, the university is responsible for curricula development, coursework, instruction, examination and grading. Moreover, the university sets the acceptance criteria into the program and the rules and regulations related to academic progression throughout the program. Further,

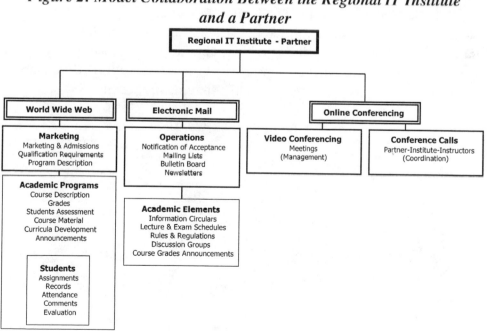

Figure 2: Model Collaboration Between the Regional IT Institute and a Partner

it is responsible for solely granting the degree upon completion by the student of all academic requirements. On the other hand, the Institute is responsible for the ICT infrastructure required, including classrooms, computer labs and the library. It is also responsible for promoting and marketing for the program, recruiting students according to the regulations set by the university, the day-to-day administration and follow-up of the program classes and students, assignment delivery and commitment to the rules and regulations of the program. With the division of duties and responsibilities, the virtual team strategy became the critical success factor for their partnership (Lipnack and Stamps, 1997). They do work across space, time and geographical boundaries with links strengthened by webs of communication technologies. Such collaborative effort is handled on a daily basis between various partners and their corresponding program coordinators. Table 1 provides statistics for the period 1994-2000 on the hybrid mode of operation between the institute and its collaborating partners.

The delivery of any of the university degree programs in Egypt with no campus and no direct contact with the granting institution was not an easy task. Moreover, a great deal of investment in infrastructure, communication, awareness, training and business process re-engineering was required and that includes both the collaborating partner and to a larger extent the Regional IT Institute. However, many lessons were learned with the implementation of the first program, which were later used to provide a model for the formulation of additional alliances with other universities and institutions in different fields to serve the Egyptian market, with value-added education representing a local capacity capable to bridge the cultural and organizational gap and adapt a program that could yield dividends in the marketplace in Egypt that brought the Regional IT Institute's market share of MBA enrolled students to almost 60% of total market share as well as having the second largest master

Table 1: Statistics 1994-2000

Online/Virtual Techniques	**92.5%**
Electronic Mail	45%
World Wide Web	42%
Video Conferencing	0.5%
Conference Calls	5%
Traditional Techniques	**7.5%**
Facsimile	4%
Snail Mail	2%
Courier	1%
Meetings	0.5%

of science in computer science in terms of enrollment and the edge of continuously introducing new programs into the marketplace since 1994, addressing specialization areas never tapped before such as globalization, e-commerce, finance and banking, and marketing management.

The first program that started in April 1994 was used as a pilot project. The project, due to the limited number of students, had to be tailored to minimize cost and rationalize the use of scarce resources. In that respect, the common semester or quarter base for the delivery of the program's 18 courses and thesis was not to be practical. Therefore, the delivery of the program that lasts for 20 months had to be engineered to run one course at a time in a sequential mode of operation. Each course was to be delivered in four consecutive weeks, three times a week in the evening from 18:00 to 22:00 with a total number of sessions of 12 for each course and a total of 48 in-class contact hours, excluding office hours which are preset with the instructor by the students. Figure 3 demonstrates the relation between the Regional IT Institute and its local students, and how it developed in terms of collaboration through a dual traditional-unconventional use of learning techniques.

This modality of operation was developed to fit the local needs for students targeting a part-time master's degree program that is also financially viable in terms of tuition and mode of payment. In that respect, it is important

Figure 3: Model Collaboration Between the Regional IT Institute and Students

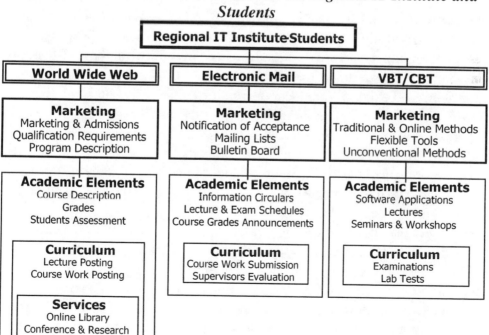

to note that the tuition of any of the Institute's collaborative degree programs does not exceed 60% of its actual tuition in its home country. Prior to the start of each course, the instructor would fly to Cairo for the delivery of the course and he/she leaves at the end of the four-week period. Week 5 is considered to be for study and Week 6 is exam week. The exam is sent electronically or by fax a few hours prior to the exam time to allow for reproduction of exam question sheets. The local program coordinator in Cairo coupled with daily correspondences with the university's headquarters via electronic mail administers Weeks 5 and 6 and manages the daily students' inquires in his/her best capacities according to the rules and regulations set by the university. For further requirements, daily exchange of messages and electronic correspondences usually occurs to provide the students an online and real-time response as much as possible. During the period of the course, students post their assignments, coursework and project work on the Web where they, through an extranet, have access to the university website with login names and personalized passwords. At the end of each course and upon the collection of all exam papers, a package is sent within 48 hours from the end of the exam via courier to the university headquarters with all answered sheets. Following the exam, a one-week break separates the following course and the cycle repeats itself until the end of the program based on the number of courses involved. Next, the thesis work starts with dual supervision from the Regional IT Institute (Cairo) and the university (Maastricht, Louisville or London) via electronic mail and weekly sessions posted via the Web. Local coaching serves for exploring issues related to methodologies and thesis writing. Finally, defending the thesis is organized in Cairo by the Regional IT Institute and attended by examiners from the collaborative university as well as other examiners and experts from local universities, research centers and institutions in Egypt to examine the students and deliver the grades. The thesis segment represents another model for virtual collaboration where supervisors are drawn globally and coordinate electronically to help students throughout the development of the thesis. Such collaboration usually entails efficiency in handling the different supervisory stages of the thesis to be able to guide the students into a well-researched and guided study.

The transfer of knowledge is similar to IT in many aspects where cultural and value-based implications vary with major impacts especially in developing countries. Therefore, the management of the knowledge transfer process has to address the local needs and the cultural values of the society to which knowledge is transferred. The experience of the Regional IT Institute implies that, in Egypt, graduate students never opt for a full-time study. Their priorities are set to find a good job rather than spend around 18 to 24 months studying. However, if arrangements are made to study outside working hours, they are more encouraged. Thus, all of the Institute's degree programs are conducted in the evenings. Finally, a major societal factor that was used in the promotion of the program was relying on the fact that students attend all their classes in Egypt without having to travel and incur extra costs. It was such cultural adaptation to the local environment and to the norms and values of the community without interfering with the academic content of the program that helped the Institute introduce and diffuse the program in the marketplace, reaching an average annual growth rate of 67% since the inception of the graduate education (degree) programs in 1994.

Managing Educational Programs Virtually

Managing a postgraduate (degree) program with an average of 35 students enrolled was a difficult task using the Regional IT Institute-partner model. The day-to-day follow-up of the operation was something new to the staff of both organizations, which entailed efficient and effective process for operations management. However, their partnership was built around collaborative learning and participation; without it neither collaboration nor learning occurs (Leidner and Javernpaa, 1995; Alavi, 1994). The partnership was formulated around jointly exerting efforts, allocating capacities, learning from past experiences, overcoming challenges and turning them into opportunities to set the rule for future growth. Figure 4 demonstrates the growth of the number of enrolled and graduating students from 1994 to December 2000.

At the organizational level, the key issue was to optimize the learning curve. Therefore, the team involved in the management and coordination of the degree programs was exposed to continuous training in management of daily operations, communication skills, planning and follow-up, crisis management, time management and customer relationship management. As for the staff, with the use of electronic mail and other IT and networking facilities, they were able to communicate and collaborate well (Schrage, 1990). The outcome of such collaboration has been proven successful with an annual growth rate of 67% in the number of students enrolled. In December 2000, there were 806 students enrolled in all education (degree) programs delivered in Cairo jointly by the Regional IT Institute through a virtual management team collaborating between Cairo, Louisville and London, that figure is expected to reach 1,000 students before the end of 2001. The Regional IT Institute education (degree) program team meets twice annually with each collaborating team from every university once in Cairo and once in its home country. The virtual management team of the Regional IT Institute is composed of six members: an academic advisor, a technical and operations

Figure 4: Regional IT Institute Degree Programs Enrolled/Graduate Students

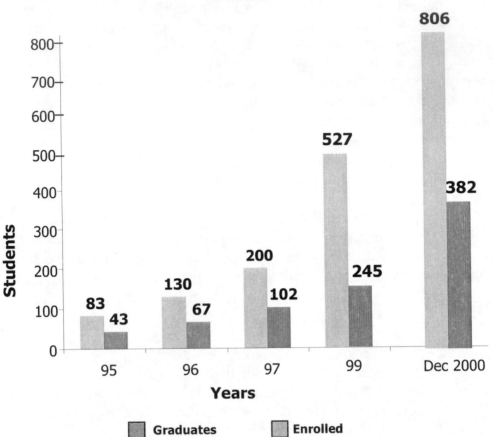

manager, marketing manager and three program coordinators, one for each collaborating university (partner). However, the use of electronic mail is intensive between two institutions on a daily basis, reaching up to 50 exchanged emails per day and more than 150 emails between the students and the university including sending assignments and inquiries. The Regional IT Institute model reflects the concept that the virtual organization can be taken to be one with a relatively small headquarters, operating with different internal units, alliances and subcontractors, in over six countries concurrently, leading to a global scale operation though built on a minor volume of resources. However, the main factors that count are accuracy, timeliness of information exchanged, and devotion and commitment of teams to realize joint success and growth. The model capitalized on the development of work processes that could maximize the internal supply chain of information and procedures within the Regional IT Institute as well as the external chain with various partners.

LESSONS LEARNED

The model of the Regional IT Institute draws a number of lessons learned that could set the stage for future reference in terms of advantages realized and disadvantages faced. The advantages include increasing competitive capability of the Institute, flexibility, greater responsiveness to market needs, improved customer services, cost benefits and improved communication and control (Grimshaw and Kwok, 1998). The use of virtual teams and advanced information and communication technology presented the Institute with a competitive advantage over its 42 rival institutions delivering similar programs in Egypt. With a 67% annual growth rate in the number of enrolled students in the MBA programs, the Institute now dominates more than 60% of the total market share in Egypt, which was only realized by the optimum investment, the use of the proper infrastructure in terms of people, information and communication. The case of the Regional IT Institute provides insights into telecommuting in the educational sector in terms of productivity gains both organizationally and individually exceeding the telecommuting statistics, showing an increase in productivity gains of 15 to 20% (Snizek, 1995).

The transfer of knowledge and the delivery of the program were not to be successfully realized without the flexibility to accommodate to local market needs in terms of domains, scheduling and administration through continuous market studies. This could be explained by the diversification that occurred to the Regional IT Institute's programs moving from one specialization in 1994 to 10 specializations in 2000, which is a reflection of the continuous monitoring of market changes and the ability to cater to these changes that helped the institute to gradually become the first choice for MBA study in Egypt. Improved customer services was a learning process that was accumulated during the past six years starting from delivering one program at a time in 1994 to 16 parallel programs in 2000. The improved service was due to the trust built between the team members, the adaptation to local conditions and the improvement of the learning process of the virtual team that was reflected in the service provided to the students (Davidow and Malone, 1992). Moreover, the use of the World Wide Web, the formulation of news groups among the students, posting of assignments on the Web and the delivery of coursework via electronic mail were adding to the virtual delivery of the program. Enrolling the students in virtual libraries was as important in providing online accessibility to a wealth of knowledge in terms of publications, references, books, journals, etc.

Cost-benefit analysis showed that the formula of the degree program was unique in terms of affordability and adaptability to market and cultural needs.

Egyptians are reluctant to travel because of cultural as well as financial reasons with a per capita of US$1,465. However, with one of the highest rates of graduate education needs in the world, Egypt lacks the supply of graduate degree programs in terms of quality and quantity with only 20 universities and a limited number of graduate placements. Respectively, many students do not have a choice but to travel, which is rather expensive and not favored. Therefore, the Institute's program provides a unique opportunity that is culturally and financially viable being delivered in Egypt and at almost 40% off the regular tuition fees in its home country. Improved internal communication and control between members of the virtual management team and the students were also among the benefits realized. The team included a heterogeneous group with diversity in cultures and nationalities. It was defined as a self-managed team with distributed expertise, however, addressing a specific organizational goal (Kristof et al., 1995). The disadvantages include cultural issues and legal problems (Grimshaw and Kwok, 1998). It was never easy to transfer the program as is and implement it successfully in the Egyptian market due to the diversity in norms and values. Therefore, adaptation of the case studies used was drawn from the local market in Egypt. Also, there were a number of problems that were addressed that relate to the administration and follow-up of the program being delivered in an unconventional way.

CONCLUSION

The Regional IT Institute is established to support the enhancement of human skills and capacities in information and communication technology aiming at the formulation of a knowledgeable information-based society. Therefore, the Institute focuses on investing in people to leverage performance and increase productivity using different information and communication technologies. This includes different modes of operation such as virtual organizations combining a hybrid of electronic and face-to-face interactions. The case of the Regional IT Institute has been built around a number of attributes including developing partnerships for knowledge diffusion to contribute to business and socioeconomic development in Egypt. The Regional IT Institute model relied on a category of learning that is technology based and where the instructor and the students are separated geographically. The model used a hybrid model of face-to-face lecturing, and virtual follow-up, study, coaching and examinations. The model provided access to knowledge for students regardless of their geographical locations while being in Egypt, using familiar technology and accommodating to different cultural norms and values. Education has always relied on the communication of thoughts and ideas between instructors and students taking many forms across different times. There has always been the barrier of geography and time. Nowadays, with the innovation in information and communication technology, these barriers are gradually being removed via information highways and advanced communication technologies (Guthrie, Olson and Schaeffer, 1998; Grimshaw and Kwok, 1998).

At the dawn of the new millennium, the megatrends driving the virtual world reflect the fact that products and services are becoming more information and knowledge-based oriented, there is an increasing societal implications of the Internet evolution, there is more globalisation of markets and resources and an increasing networking capacities, enabling collaboration, responsiveness and flexibility (Skyrme, 1998). The case of the Regional IT Institute as a hybrid model for virtual organizations shows that the world is rapidly moving away from the belief that there has to be one theory of the organization and one ideal structure. It demonstrates that the options are wide open to adapt organizational development and knowledge management to fit local market needs and conditions (Drucker, 1997).

REFERENCES

Alavi, M. (1994). Computer-mediated collaborative learning: An empirical evaluation. *MIS Quarterly*, 18(2), 159-174.

Appel, W. and Behr, R. (1997). Towards the theory of virtual organizations: A description of their formation and figure. *Virtual Organization Net Newsletter*, 2(2).

Brigham, M. and Corbett, M. (1996). Trust and the virtual organization: Handy cyberias. In Jackson, P. and Wielen, J. V. D. (Eds.), *Proceedings of the Workshop "New International Perspectives on Telework: From Telecommuting to the Virtual Organization,"* Brunel University, West London, United Kingdom, July-August, 1.

Byrne, J. A., Brandt, R. and Port, O. (1993). The virtual corporation: The company of the future will be the ultimate in adaptability. *International Business Week*, February, 36-40.

Coates, J. F. (1994). Managing scientists in the virtual corporation. *Research Technology Management*, November-December, 37, pp. 6-8.

Davidow, W. H. and Malone, M. S. (1993). *The Virtual Corporation*. Harper Collins.

Drucker, P. F. (1997). Introduction: Toward the new organization. In Hessekbein, F., Goldsmith, M. and Beckhard, R. (Eds.), *The Organization of the Future*, 1-5, San Fransisco, CA: Jossey-Bass Publishers.

Goldman, S. L., Nagel, R. N. and Preiss, K. (1995). *Agile Competitors and Virtual Organizations: Strategies for Enriching the Customer*, New York, NY: Van Nostrand Reinhold.

Gray, P. (1995). The virtual workplace. *ORMS Today*, A publication of INFORMS, August.

Grenier, R. and Metes, G. (1996). *Going Virtual: Moving Your Organization into the 21st Century*, Prentice Hall.

Grimshaw, D. J. and Kwok, F. T. S. (1998). The business benefits of the virtual organization. In Igbaria, M. and Tan, M. (Eds.), *The Virtual Workplace*, Hershey, PA: Idea Group Publishing.

Gristock, J. J. (1997). The combinationary role of virtual experiences: Implications for knowledge exchange. *Virtual-Organization Net Newsletter*, 2(2).

Guthrie, R., Olson, P. C. and Schaeffer, D. M. (1998). The professor as teleworker. In Igbaria, M. and Tan, M. (Eds.), *The Virtual Workplace*, Hershey, PA: Idea Group Publishing.

Kamel, S. (1998a). Humanware investment in Egypt. *Proceedings of the International Federation for Information Processing WG9.4 Working Conference on Implementation and Evaluation of Information Systems in Developing Countries*, Asian Institute of Technology, Bangkok, Thailand, 18-20 February.

Kamel, S. (1998b). IT diffusion through education and training. *Proceedings of the 8th Annual BIT Conference on Business Information Management-Adaptive Futures*, 4-5 November, Manchester, United Kingdom.

Kamel, S. (1999). Web-based interactive learning. *Information Management Journal*, Spring, 12(1-2), 6-19. Retrieved on the World Wide Web: http://www.idea-group.com/im99-1.htm.

Knoll, K. and Jarvenpaa, S. L. (1998). Working together in global virtual teams. In Igbaria, M. and Tan, M. (Eds.), *The Virtual Workplace*, Hershey: Idea Group Publishing.

Kocian, C. (1997). The virtual center: A networking cooperation model for small businesses. *Virtual-Organization Net Newsletter*, 1(2).

Kristof, A. L., Brown, K. G., Sims Jr., H. P. and Smith, K. A. (1995). The virtual team: A case study and inductive model. In Beyerlein, M. M., Johnson, D. A. and Beyerlein, S. T. (Eds.), *Advances in Interdisciplinary Studies of Work Teams: Knowledge Work in Teams*, 2, 229-253, Greenwich, CT: JAI Press.

Leidner, D. E. and Jarvenpaa, S. L. (1995). The use of information technology to enhance management school education: A theoretical view. *MIS Quarterly*, 19(3), 265-291.

Lipnack, J. and Stamps, J. (1997). *Virtual Teams*. New York, NY: John Wiley & Sons, Inc.

Nohria, N. and Eccles, R. (1992). Face-to-face: Making network organizations work. In Nohria, N. and Eccles, R. G. (Eds.), *Networks and Organizations: Structure, Form and Action*, Boston, MA: Harvard Business School Press, 288-308.

Palmer, J. W. (1998). The use of information technology in virtual organizations. In Igbaria, M. and Tan, M. (Eds.), *The Virtual Workplace*, Hershey, PA: Idea Group Publishing.

Regional IT Institute. (2001). Retrieved on the World Wide Web: http://www.riti.org.

Rockart, J. F. and Short, J. E. (1991). The networked organization and the management of interdependence. In Morton, M. S. (Ed.), *The Corporation of the 1990s*, 189-219, New York.

Schrage, M. (1990). *Shared Minds*, New York, NY: Random House.

Sieber, P. (1997). Virtual organizations: Static and dynamic viewpoints. In Griese, J. and Sieber, P. (Eds.), *Virtual-Organization Net Newsletter*, 1(2). Department of Information Management, University of Berne, Retrieved on the World Wide Web: http://www.virtual-organization.net/news/nl_1.2/sieber.html.

Skyrme, D. J. (1998). The realities of virtuality. In Sieber, P. and Griese, J. (Eds.), *Proceedings of the VoNet Workshop*.

Snizek, W. E. (1995) Virtual offices: Some neglected considerations. *Communications of the ACM,* (September) 38, 15-17.

Strausak, N. (1998). Resumee of votalk. In Sieber, P. and Griese, J. (Eds.), *Proceedings of the VoNet Workshop*.

Ungson, G. R. and Tundel, J. D. (1998). *Energy of Prosperity: Templates from the Information Age*, London, UK: Imperial College Press.

Wolff, M. (1995). *Ki-Net New Organizational Structures for Engineering Design*, March. Retrieved on the World Wide Web: http://www.ki-net.co.uk/ki-net/part1.html.

BIOGRAPHICAL SKETCH

Sherif Kamel is a Visiting Assistant Professor at the American University in Cairo in the Management Department at the School of Business, Economics and Communication. He is also the director of the Regional IT Institute (Egypt). He has published more than55 articles in IT transfer to developing countries, e-commerce and human resources development. He designs and delivers professional development programs in information systems applications for public and private-sector organizations in different countries in Africa, Asia, the Far East and Eastern Europe. He is one of the founding members of the Internet Society of Egypt. He serves on the editorial and review boards of a number of information systems and management journals. He is the Associate Editor of the Annals of Cases on Information Technology Applications and Management in Organizations *and is currently the VP of Communications for the Information Resources Management Association (IRMA).*

Section III

Research Cases

IT Industry Success in Small Countries: The Cases of Finland and New Zealand

Rebecca Watson
PA Consulting Group, Australia

Michael D. Myers
University of Auckland, New Zealand

EXECUTIVE SUMMARY

Given the importance of the information technology industry in today's global economy, much recent research has focused on the relative success of small countries in fostering IT industries. This case examines the factors of IT industry success in small developed countries, and compares two such countries, Finland and New Zealand. Finland and New Zealand are alike in many respects, yet Finland's IT industry is more successful than New Zealand's. Three major factors that impact on the development of a successful IT industry are identified: the extent of government IT promotion, the level of research and development, and the existence of an education system that produces IT-literate graduates.

INTRODUCTION

At the beginning of the 21st century, information and communication technologies are creating global markets for goods and services. These technologies are impacting on every aspect of our lives, including how people work, communicate and entertain themselves. Many economists have started to suggest that we may be entering a new era of greater productivity (without inflation) in the "knowledge economy" of the future.

What is not clear, however, is how many countries will be able to adapt and develop new information-based industries of their own. In this new global, knowledge-based economy of the future, it is likely that some countries will thrive and become significant players, while others will not. Those countries that cannot adapt will suffer and may find themselves as producers of low-value products for wealthier nations.

Given the importance of the IT industry in today's global economy, much recent research has focused on the relative success of small countries in fostering IT industries. This chapter builds on this earlier work and examines the factors of IT industry success in small developed countries, focusing on two such countries, Finland and New Zealand. We chose Finland and New Zealand for this study because they share many common characteristics, yet Finland's IT industry is more successful than New Zealand's (as will be shown below). Using a modified version of the theoretical model suggested by Ein-Dor et al. (1997), this chapter suggests factors that may contribute to the differing levels of IT industry success. For the purposes of this study, we define IT as computer hardware and software but exclude embedded hardware and software in other products (e.g., washing machines).

With respect to the static or snapshot nature of the data presented here, Kraemer, Gurbaxani and King (1992) argue that the

> *. . . relationship between interventions and their consequences is best revealed by careful, longitudinal study that links together specific policies and actions with particular results. Such study is badly needed and in some limited instances has begun, but to date the best assessments are limited to cross-sectional evaluations of correspondences between policies and economic measures of computer-related activities in given countries* (1992, p. 149).

We agree with them that, in the long run, a longitudinal analysis is undoubtedly better for many of the things we are trying to find out. However, the objective of this first study is an exploratory review benchmarking the state of the IT industries in the two countries. We believe that a cross-sectional evaluation is sufficient for this purpose.

The chapter proceeds as follows. In the following section the theoretical framework is presented. The research methodology is then described. This is then followed by an analysis of the data relating to the two countries. In this particular section, each country is briefly described and the controlled variables are presented. Then the dependent variables relating to IT industry success are evaluated. The main part of the analysis describes the exogenous and endogenous factors for each country and, wherever possible, an attempt is made to relate these variables to the differing degrees of IT industry success. The final section is the conclusion.

THEORETICAL FRAMEWORK

Most of the previous research in this area has compared a reasonably large number of countries. For example, Blanning et al. (1997) examined the information infrastructure of 12 Asia Pacific nations; Dedrick et al. (1995) examined reasons for the success of IT industries in nine small countries from around the world; and Kraemer, Gurbaxani and King (1992) discussed the diffusion of computing use in nine Asia Pacific nations. Generally, these studies have examined a small range of factors that impact on either the success of a nation's IT industry or its extent of IT usage.

In contrast, Ein-dor et al. (1997) examined only three small countries—Israel, New Zealand and Singapore. These three countries were of similar size and economic development, however, they were experiencing differing levels of IT industry success. They were examined in depth to determine those factors that impacted on the development of a successful IT industry in a small country. Ein-dor et al.'s (1997) study has been one of the few pieces of research that has examined only a small number of countries in an in-depth manner. This research study adopts the approach and model as used by Ein-dor et al. (1997) and compares just two countries, New Zealand and Finland.

Ein-dor et al.'s (1997) model was largely based on Grossman and Helpman's (1991) macroeconomic theory concerning the relationship between technology development, trade, and growth as applied to small open economies. The latter suggested that "growth stems from endogenous technological progress, as farsighted entrepreneurs introduce innovative (intermediate) products whenever the present value of the stream of operating profits covers the cost of product development." Grossman and Helpman (1991) postulated that the best growth path can be attained with subsidies to both R & D and the production of "intermediates" (those products which are used to produce consumer goods). The second-best growth path can be achieved with subsidies to R&D alone.

In order to study IT industry success in accordance with the above theory, Ein-dor et al. (1997) considered four groups of variables. These variables are all frequently quoted in the context of industrial success. The variables they considered were as follows:

1. *Controlled variables*—country size and economic development.
2. *Dependent variables*—those which define IT industry success.

3. *Exogenous factors*.
4. *Endogenous factors*:
- Domestic IT use.
- Firm strategies.
- Government IT policies.
- Government education policies.

In our study, we considered the same four groups of variables, however, we decided to replace "firm strategies" with "level of research and development," because the latter appears to have more explanatory power. The major factors that we considered in our study are represented graphically in Figure 1.

This model was then used to compare and explain IT industry success in Finland and New Zealand. More detail concerning the original model can be found in Ein-Dor et al. (1997).

METHODOLOGY

The research methodology involved collecting a range of quantitative and qualitative data about Finland and New Zealand as suggested by the theoretical model, with an attempt to provide as much comparability as possible. Wherever possible, data on the two countries were taken from common sources. The data are presented according to the factors suggested by Figure 1. Data for this research was collected from a variety of sources including OECD reports, official government publications, industry surveys, newspaper articles, websites and international research companies. All monetary figures used in this chapter have been converted to U.S. dollars.

COUNTRIES' ANALYSIS

The structure of this next section is as follows. First, each country is briefly described and the controlled variables are presented. Then the dependent variables relating to IT industry success are presented. The main part of the analysis describes the exogenous and endogenous factors for each country

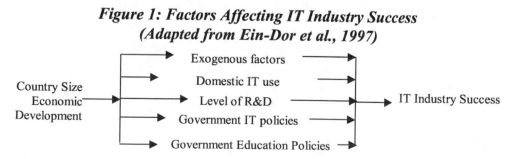

Figure 1: Factors Affecting IT Industry Success
(Adapted from Ein-Dor et al., 1997)

and, wherever possible, an attempt is made to relate the variables to the differing degrees of IT industry success. At the end of each section a brief summary is provided.

Controlled Variables-Country Comparison

Finland and New Zealand share many similar characteristics with regard to country size and economic development (see Table 1).

Country Size

Dedrick, Goodman and Kraemer (1995) define a small country as one which has fewer than 10 million people. Finland and New Zealand, with populations of 5.1 million and 3.6 million respectively, are thus considered small countries. The physical size of the two countries is also similar which, along with a similar population, means that their population densities are almost identical.

Economic Development

In the early 19th century, Finland was one of the poorest and most agrarian areas in Europe. Because of the cold climate, agricultural development was limited in most areas except for forestry (Lyytinen & Goodman, 1999). The economy made a turnaround in World War II (WWII) when Finland developed a machine industry. In the 50 years since WWII, Finland successfully changed gear and cut its reliance on producing primary commodities (Information Technology Advisory Group, 1999).

Today, Finland is considered a highly industrialized, largely free market economy. Finland's traditional forestry industry accounts for less than 3% of GDP, despite experiencing growth output. High technology industries account for a significant percentage of the Finnish GDP. Finland has thus transformed itself from a commodity-based economy to one that has embraced new technologies (Information Technology Advisory Group, 1999). As a result, Finland's GNP per-capita has grown to more than $24,000, placing it among the world's wealthiest countries.

New Zealand has historically relied heavily on agriculture. This mainstay of the New Zealand economy went through a massive growth period in the late 1800s due to the invention of refrigerated ships. These new ships enabled New Zealand meats and dairy products to reach Britain, which was New Zealand's primary market at that time (Myers, 1996).

New Zealand depended heavily on primary commodity exports to Britain up until the end of World War II. In the early 1980s the government initiated major economic restructuring in order to move from an agrarian economy to an industrialised, free-market economy that could compete globally. This dynamic growth boosted real incomes, strengthened business confidence and increased demand for New Zealand exports in the Asia Pacific region (Central Intelligence Agency, 1999). Exports also diversified away from dairy, meat and wool into such industries as forestry, horticulture, fishing and manufacturing (Myers, 1996).

Today, New Zealand still relies heavily on external demand for its agricultural-based products. However, New Zealand's modern agriculture, forestry and fishery industries produce a large variety of products with added value, catering for niche markets in many countries (Ministry of Commerce, 1996). While these new value-added products have been contributing to New Zealand's GNP per-capita, it has not been experiencing the growth of countries such as Finland.

Table 1: Country Size and Economic Development (1998)

	Finland	New Zealand
Country Size		
Population (millions)	5.1	3.6
Population Rank	105	120
Surface Area (km^2)	337,030	268,680
Population Density (people per square kilometre)	15	13
Economic Development		
Per-capita GNP -1997	24,790	15,830
Per-capita GNP Rank -1997	20	35
Per-capita GNP Average Annual Growth 96-97	5.9%	-0.4%
Per-capita GDP	22,121	15,470
Per-capita GDP Rank	16/53	23/53
Overall Productivity (GDP per person employed)	48,319	38,305

Sources: Central Intelligence Agency, 1999; IMD, 1999; World Bank, 1999; World Economic Forum, 1998

Summary

Finland and New Zealand are remarkably alike in terms of country size and economic development. Both Finland and New Zealand entered the 20th century with a heavy dependence on commodity products. Forestry has played the same role in Finland's development as agriculture has played in New Zealand's development. According to our research, the only difference between the two countries is that Finland has moved away from its dependence on forestry and has grasped on to new technologies, whereas agriculture is still a major part of New Zealand's economy.

Dependent Variables—IT Industry Development and Success

Table 2 summarizes the indicators of IT production and development. As can be seen, Finland's IT industry is more successful than New Zealand's.

IT Industry Sales

The IT industry is a very important sector of the Finnish economy, with sales accounting for around 9% of Finland's GDP. Sales in the New Zealand IT market are substantially smaller than in Finland, at around 4% of GDP.

Number and Size of IT Firms

It is estimated that there are about 4,200 IT firms in Finland. The majority of these firms are very small, with 3,700 of them having fewer than five employees (Nygard & Kunnas, 1998). Four IT firms are included in Finland's top 50 companies, with Nokia being the largest company in Finland. Three out of four of these companies sell telecommunications products and services.

New Zealand has just over 2,500 firms in the IT industry, with the majority of firms being in the software or services sector. Like Finland, most of these firms are very small. Telecom New Zealand, the second largest company in the country, is the only IT firm included in New Zealand's top 50 companies. However, there are 13 IT firms listed in the top 200 companies (Deloitte & Touche Consulting Group, 1998).

IT Employment

In Finland, 5.5% of persons employed in the private and public domains during 1997 were engaged in jobs in the IT industry. Because of the growth of the IT industry, it accounted for almost 13% of the total increase in the number of employed persons in 1996-1997 (Statistics Finland, 1999a).

In New Zealand, the number of people who were working in the IT industry in 1996 increased by 4% to 41,823. However, as a percentage of the working population, the number employed in the IT industry decreased from 2.9% in 1991 to 2.6% (March, 1999).

Table 2: IT Industry Success (1998)

	Finland	New Zealand
IT Industry Development		
IT Industry Sales (billions)	11,087	2,155
IT Sales/GDP	9.0%	4.1%
Number of Firms in IT	4,200	2,500
IT Firms/Largest firms	4/50	1/50
IT Industry Employment	5.5% (1997)	2.6% (1996)
IT Industry Success		
IT Exports:		
Hardware (millions)	7,255	175
Software (millions)	488	123
Total (millions)	7,743	298
High Technology Exports (millions) –1997	8,797	428
High Technology Exports/Manufactured exports -1997	26%	11%
Stock Market Listings:		
Domestically listed IT firms	27	3
Internationally listed IT firms	1	0

Sources: Computerworld, 1999; Deloitte & Touche Consulting Group., 1998; Ein-Dor et al., 1997; Finnfacts, 1999; Helsinki Stock Exchange, 1999; March, 1999; Nygard & Kunnas, 1998; Statistics Finland, 1999a; Statistics Finland, 1999b; World Bank, 1999

IT Exports

Finland's IT exports have increased rapidly in recent years. Strong export positions have been created in mobile phones, personal computers and software products. In fact, Finland and Ireland are the only two European countries with positive trade balances for IT products and services. Moreover, exports of Finnish IT products and services are expected to double by 2002 (Lyytinen & Goodman, 1999).

Much of this rise in Finnish IT exports has been thanks to global telecommunications company Nokia, which to some has become better known than Finland itself. Nokia's success is partly attributable to its acceptance of the Global System for Mobile Communications (GSM) in 1989, which made Finland the first country to launch a digital mobile network. GSM has since become the standard in all other continents except the Americas (Lyytinen & Goodman, 1999). Nokia's growth has been such that if it continues at the same pace, its revenues will exceed the budget of Finland some time early next decade (International Telecommunication Union, 1999).

New Zealand's export sector has failed to develop an information technology segment similar to that of Finland (Information Technology Advisory Group, 1999). New Zealand's IT hardware exports represent a fraction of Finnish hardware exports. However, hardware exports are increasing and over the last five years have experienced an average annual growth rate of 25%. Telecommunications hardware has been one of the major contributors to increases in hardware exports. Exports of software and services increased by 56% from 1997 to 1998 (March, 1999).

High Technology Exports

In 1998, nearly a fifth of Finland's exports were high-technology products, up from only 4% a decade ago (Edmondson, 1995). In 1995, the value of high-technology exports exceeded that of imports for the first time. Further, much of Finland's high-technology production is oriented to foreign markets, as indicated by the fact that 25% of manufactured exports are in high-technology sectors (Kraemer & Dedrick, 1992).

Like IT exports, high technology exports in New Zealand are significantly smaller than in Finland. At $428 million, New Zealand exports are equivalent to 5% of Finnish high-technology exports. Also, high-technology production accounts for only 11% of total manufactured exports.

Stock Listings

There are 150 firms listed on the Helsinki Stock Exchange. Twenty-seven of these firms are in the IT industry, which represents 18% of all firms on the stock exchange. One Finnish firm is listed on international stock exchanges. This is of course Nokia, which is listed on six exchanges including the New York Stock Exchange (Perry, 1999).

New Zealand's stock exchange, the NZSE40, only lists the top 40 public companies. With three of these companies in IT, the industry represents 7.5% of all listed companies.

Summary

Finland's IT industry is far more successful than New Zealand's. At this stage, most of Finland's rapid rise to success in IT can be attributed to Nokia. Even so, the fact that such a small country can spawn a multi-billion-dollar company is a notable achievement. The remaining sections of this chapter present the three factors that help explain the differing levels of IT success exhibited by the two countries.

Exogenous Factors

Table 3 presents data on the exogenous factors, i.e., geographical location, proximity to major IT markets, climate and national culture.

There are clear differences between Finland and New Zealand in terms of the exogenous factors presented in Table 3, however we believe it is unlikely that any of them have a significant long-term impact on IT industry success. This view is supported by Ein-dor et al. (1997), who suggest that geographical, cultural and environmental differences do not seem to explain differential levels of IT industry development. We suggest that location may have a slight impact on hardware industry success, but is mostly irrelevant for software industry success.

Endogenous Factors

Domestic IT Use

The effective use of technology has been identified as an important component of industrial growth (Kraemer et al., 1992). Ein-dor et al. (1997) agree with this proposition and suggest that "a high level of local IT use is a necessary but not a sufficient condition for IT industry development." Table 4 presents data on the domestic use of IT.

As can be seen, both Finland and Zealand are extensive users of IT, as measured by telecommunications development, Internet use and IT spending. Given the overall similarity in domestic IT use, this factor does not provide an explanation for Finland's more successful IT industry. This finding supports the earlier study of Ein-dor et al. (1997) and thus we conclude that a high level of domestic IT use, though important, is not sufficient to explain IT industry success.

Level of Research & Development

Investment in research and development (R&D) fuels innovation and raises the technological sophistication of manufacturing processes. The amount of R&D undertaken in an industry is often an indication of potential growth, productivity gains and competitive advantage (Dobbeck & Woods, 1994). Table

Table 3: Location-Related Factors

	Finland		New Zealand	
Geographical Location				
Physical Location	North Europe		South Pacific	
Distance from nearest neighbour	0km (Sweden, Norway & Russia)		1600km (Australia)	
IT Market Proximity	*Distance*	*Trade*	*Distance*	*Trade*
Europe	Close	Most	Far	Middle
Far East/South West Pacific	Far	Least	Medium	Most
United States	Medium	Middle	Far	Least
Climate				
Type of Climate	Cold temperate/Subartic		Temperate/Subtropical	
Mean Annual Temperature	5.3° Celsius		14° Celsius	
Highest Temperatures	30° Celsius		30° Celsius	
Lowest Temperatures	-50° Celsius		-20° Celsius	
National Culture				
Ethnic Groups	Fin 93%, Swede 6%, Lapp 0.11%, Gypsy 0.12%, Tatar 0.02%		European 79.1%, Maori 9.7%, Pacific Islander 3.8%, Asian and others 7.4%	
Official Languages	Finnish, Swedish		English, Maori	
Life Expectancy	74 male, 77 female		74 male, 80 female	
Infant Mortality/1000 live births	7		5	
Literacy	100%		100%	
Newspaper Circulation	524/1000 people		304/1000 people	

Sources: Central Intelligence Agency, 1999; Virtual Finland, 1999; World Almanac, 1997

Table 4: Domestic IT Use (1998)

	Finland	New Zealand
Telecommunications Development		
Telephones Lines per 100 people	55.39	47.91
Mobile Phones per 100 people	57.18	20.26
Fax Machines per 100 people -1997	3.85	1.79
Telecommunications Revenues (billions)	3.6	1.8
Per-Capita Telecommunications Revenues	706	500
Internet Use		
Internet Hosts per 100 people	10.4	5.2
Internet Hosts per 100 people World Rank	1	7
Web Server Sites per 100 people World Rank	15	10
IT Spending		
Total IT Spending (billions)	3.34	2.14
IT Spending/GDP	2.7%	4.1%
Personal Computers per 100 people -1997	31.1	26.4

Sources: IMD, 1999; Information Technology Advisory Group, 1999; Inter national Data Corporation, 1999; International Telecommunication Union, 1999; Minges, 1999; OECD, 1998a; World Bank, 1999

5 presents a range of data that measures the extent of R&D in each country. Total R&D expenditures, sources of R&D funding, and the number of scientists and engineers are included because these factors are believed to be associated with success in IT production (Kraemer & Dedrick, 1992).

Extent of R&D

Research and technological development have been priority areas in Finland for 20 years. As a result, investments in R&D have risen steadily for the past 15 years. The growth in research spending has been among the highest in the OECD countries, at around 16% per annum. The proportion of GDP spent on R&D has risen from 1.8% in 1987 to 2.7% in 1997 (Statistics Finland, 1999b).

Table 5: Level of Research and Development

	Finland	New Zealand
Extent of R&D		
Gross Domestic Expenditure on R&D (millions)	2,150	589
Per-capita R&D Expenditure	421.0	164.4
R&D as a percentage of GDP –1997	2.71	0.98
Percentage of R&D Funded by the Private Sector	59.5	33.7
Percentage of R&D Funded by the Public Sector	35.1	52.3
R&D Personnel per 1000 workers	13.3	6.1
Scientists and Engineers in R&D per 1,000,000 people - 1996	3,675	1,778
Performance of R&D		
Percentage of R&D Performed by the Private Sector	63.2	27.0
Percentage of R&D Performed by the Public Sector	36.1	72.9
R&D Tax Incentives		
B-Index 1996– for every 1$ invested it costs:	1.01	1.13
B-Index Rank in the OECD Countries –1996	16	25
R&D Outcomes		
National Patent Applications –1996	64,818	28,368
Resident Patent Applications -1996	3,262	1,421

Sources: (Finnfacts, 1999; Information Technology Advisory Group, 1999; OECD, 1998b; World Bank, 1999).

Research activity in IT has also been steadily growing over the last decade. Much of the initial increase in IT R&D was due to the Finnish government increasing and reallocating funding to this area. The Finns have now achieved excellent results in areas including neural computing, telecommunications protocols, databases, information systems and software engineering (Lyytinen & Goodman, 1999).

The majority of Finland's R&D spending is funded by the private sector. This trend has been increasing, and in the past few years Finnish companies have raised their R&D investments by about 15 to 20% annually (Maenpaa, 1999). The reason for this increase is that the government has been actively stimulating R&D spending in industry in order to decrease its own R&D spending. While the world's top 300 companies spent an average of 4.6% of sales on R&D, Finnish companies spent more than double the OECD average, at 10.4%. Nokia alone spent more on R&D than the whole of New Zealand (Information Technology Advisory Group, 1999).

In terms of total R&D expenditure, New Zealand is more than halfway down the OECD country list, with only 0.98% of GDP going into R&D (ITAG, 1999). Though efforts are being made to increase total R&D spending, these efforts are insignificant and are unlikely to make a major difference to R&D as a percentage of GDP (OECD, 1996c). With total R&D expenditures low, it is not surprising that IT R&D spending is dismal. In 1995, a mere $2.86 million of government funding was spent on IT R&D. Total private and public sector investment in IT R&D accounted for only 1.3% of total R&D spending (Information Technology Advisory Group, 1999).

The most significant contributor to this small amount of R&D is New Zealand's government. Even so, only 0.61% of GDP goes into government-funded R&D, less than half the OECD average (Information Technology Advisory Group, 1999). Because the majority of the research is government funded, R&D is currently disproportionately skewed towards the agriculture sector. The government's main science fund, the Public Good Science Fund, concentrates on the horticulture, marine and forestry sectors. Private sector investment is almost insignificant and has declined in recent years (OECD, 1996c). This is particularly unfortunate for the IT industry, which receives little government support and is thus predominantly funded by the private sector. With decreasing private sector R&D investment, it is widely agreed that there needs to be a much greater investment by the public sector in IT-related R&D (Ministry of Commerce, 1999).

Performance of R&D

The majority of Finland's R&D is performed by the private sector, which ensures that research is directed toward commercially viable areas. This large amount of private-sector R&D is due to work of the government-funded Technology Development Center of Finland (TEKES), which has been fostering industry-oriented R&D since 1983. TEKES created a tradition and mode of close industry-university interaction more advanced than most other countries. As a result, R&D and innovations have been encouraged in many industries, including the growing IT industry. TEKES also supports companies in their risk-bearing R&D projects with grants and soft loans (Lyytinen & Goodman, 1999).

In contrast to Finland, most of New Zealand's R&D is performed by the public sector, specifically by nine state-owned research companies, called Crown Research Institutes (OECD, 1996c). These companies tend to concentrate on basic research from which commercial applications can be derived (Ministry of Commerce, 1993). They also tend to focus on primary production industries, rather than IT. This tendency to research in traditional industries is likely to change, as a recent alteration to government policy has opened up Crown Research Institute funding to companies and researchers. This change should increase the amount of R&D that is being performed by the IT industry.

R&D Tax Incentives

The Finnish government encouraged private R&D investment by providing tax incentives. Although these tax incentives have been discarded, they did stimulate R&D when they were introduced (OECD, 1996b).The current tax situation is still favourable to R&D. Finnish companies can fully deduct current business expenditures on R&D in the year incurred, machinery and equipment can be deducted fully in the year incurred, and buildings for research purposes may be depreciated in Finland at 20% per year (OECD, 1996a).

New Zealand's R&D tax situation is the least favourable of the OECD countries. No tax incentives are offered, and for every dollar a private company invests in R&D, it costs them $1.13 (Information Technology Advisory Group, 1999). New Zealand is also the only OECD country that does not allow current business expenditures on R&D to be fully deducted in the year incurred. This is because the tax law maintains that any kind of R&D expenditure is an investment expense and needs to be capitalised accordingly (OECD, 1996a). The requirement to capitalise R&D investment may lead to an under-reporting of R&D, or it may act as a disincentive to such investments (Information Technology Advisory Group, 1999). Even though it has been found that short-term tax breaks can help stimulate R&D spending, the former National government of New Zealand opposed allowing any such special treatment to move into the tax system (Myers, 1996). New Zealand is certainly not encouraging increased private-sector investment in R&D.

Summary

Finland and New Zealand have vastly different levels of R&D investment. Finland spends an increasingly significant proportion of its GDP on R&D. Most of this R&D is funded and performed by the private sector. This ensures that it is concentrated in growth industries such as IT. Further, R&D continues to be stimulated through favourable tax conditions and grants from TEKES. In New Zealand, R&D spending is extremely low. The government funds and performs most of the research, which tends to be concentrated in agricultural industries. As a result, R&D in the IT industry is minimal. To make matters worse, New Zealand's tax situation does not encourage increased private-sector R&D. Therefore, a country's level of R&D appears to affect the development of a successful IT industry. Moreover, a high level of private-sector R&D investment seems to be important for IT industry success.

Government IT Policies

Much research has been devoted to determining what role the government should play in the development of an IT industry. The majority of this research has concluded that direct government promotion seems necessary for the development of a successful IT industry (Dedrick et al., 1995; Ein-Dor et al., 1997; Kraemer & Dedrick, 1992; Kraemer et al., 1992). Therefore, this section examines the extent of government promotion in both Finland and New Zealand. The focus here is on the existence of national IT strategies, the priorities attached to IT and the role of government IT organizations. Table 6 presents a summary of government IT policies in the two countries.

National IT Strategies

For a long time Finland has been seeking to play a pioneering role in implementing an information society. In order to do this, Finland's Information Technology Advisory Board (1976 to 1991) deemed that a national information society strategy was necessary. This idea was supported by a country review of Finland's IT and telecommunications policies performed by the OECD in 1990 to 1992. The OECD country review concluded that while Finland had reached an astonishingly high level of IT and telecommunications penetration, the country lacked a clear statement of strategy in these areas. Consequently, the Ministry of Finance was given the task of preparing

Table 6: Government Promotion of IT

	Finland	New Zealand
National IT Strategies		
IT Strategy	Yes	No
Type of Promotion of IT Use	National Strategy	Economic Policy
Type of Promotion of IT Development	National Strategy	None
IT Priority		
IT Use Priority	High	Low
IT Development Priority	High	Low
Government IT Organisations		
Number of Policy Setting Organisations	1	0
Number of Advisory Organisations	3	3

one. The report, entitled "Finland Towards the Information Society—A National Strategy," said that Finland should aim to use IT to gain and maintain a competitive edge within the world economy (Ministry of Finance, 1996).

Finland's government promotes the utilization of information networks and the Internet. The government ensures that basic information society skills are available to all. It is also in the process of developing an Information Highway that will eventually reach homes, public services and small and medium-sized enterprises (Ministry of Finance, 1996).

The Finnish government is actively promoting the development of the IT industry. Programs and grants have been established that seek to strengthen the competitiveness of the industry, to create new products, new businesses and new jobs (Ministry of Finance, 1996). Further, the government assisted in the creation of Oulu's Technopolis, the world's northernmost science park. Technopolis is home to the world's best telecommunications and electronics technology and to more than 100 new technology ventures (Edmondson, 1995).

Unlike Finland, New Zealand has never had a formal IT strategy, nor are there any plans to implement one. The previous National government rejected the idea of a formal IT strategy because it preferred to let the free market reign. The government believed that support for the IT industry would go against its philosophy of deregulation and would foster a return to the protectionist welfare state of the early 1980s. It also feared that by directly supporting one industry, other industries would be penalized. As such, all industries are supported through wider economic strategies. For example, the Asia 2000 strategy aims to create favourable conditions for New Zealand exporters to move into Asian markets. Further, the government's macroeconomic policies aim to keep inflation low and to maintain a favourable exchange rate (Myers, 1996).

In terms of IT development, government promotion is minimal. There are no special tax incentives, and few loans or grants are available for IT companies (Ein-Dor et al., 1997). The government also does not insist on purchasing local IT products, even though such decisions can have a tremendous impact on industry development. In one break from its free-market stance, the government worked with local businesses to create the Canterbury Technology Park. This park was developed to enable high-technology companies to interact with local academic and research institutions (Kraemer & Dedrick, 1993).

IT Priority

The development and use of IT receives a very high priority in Finland. The fact that Finland created a national IT strategy highlights the importance attached to IT. Also, parallel IT strategy work has been going on in other areas in Finland, most notably in the industrial, educational, cultural, and health and welfare sectors (Ministry of Finance, 1996).

The IT priority in New Zealand remains low. However, a relatively new organization, the Information Technology Advisory Group (ITAG), was established to provide policy advice to government on IT.

Government IT Organizations

Finland has one government IT organization (the Technology Department) that is responsible for developing IT-related strategies and policies. Three more government IT organizations provide advice and stimulate discussion on IT-related issues. One of these, the Science and Technology Policy Council, discusses important questions relating to the advancement of science, technology and scientific education. The other two include the National Information Society Forum and a Government Committee for Information Society Issues (Ministry of Finance, 1996).

The Finnish government has also promoted the development of inter-ministerial clusters. These clusters bring together technology developers, public service providers and policy makers. The clusters partake in IT related policymaking, research and technology development in an interactive way that creates a fertile ground for innovation (Maenpaa, 1999).

No New Zealand government IT organization is responsible for setting IT-related policies. However, the government does receive advice on IT issues from three main sources. The first is the Information Technology Policy Unit, which resides within the Ministry of Commerce. The purpose of the unit is to provide economic policy advice. Further sources of advice include ITAG and the Electronic Commerce Steering Committee (Information Technology Advisory Group, 1999). All three of these organizations have been established within the last six years.

Summary

Finland and New Zealand are clearly different in terms of government IT promotion. Finland has a high-priority national IT strategy that promotes both IT use and industry development. As a result, Finland is both a heavy user of IT and a significant player in the international IT industry. New Zealand has no IT strategy and IT is treated in the same manner as all other industries. While this lack of government support has not hindered New Zealand's adoption of IT, it has hindered its IT industry success. Government promotion of IT is thus one factor that impacts on the development of a successful IT industry.

Government Education Policies

Total Education Expenditure

It is believed that a top-quality education system is essential to being successful in the information age. Both Finland and New Zealand consider education important. When compared to other OECD countries, they exhibit among the highest education expenditures as a percentage of GDP. Though Finland spends slightly more of its GDP on education than New Zealand, both countries have well-developed education systems (Butler & Zwimpfer, 1997). Finland and New Zealand's education systems are compared in Table 7.

School Education

The Finnish education system has actively promoted IT skills, resulting in extensive IT usage throughout Finland's schools. Primary and secondary

Table 7: Government Education Policies

	Finland	New Zealand
Total Education Expenditure		
Public Expenditure on Education/GDP -1995	7.6%	7.3%
Annual Expenditure per student relative to per-capita GDP –1993	20.53	15.30
School Education		
School Education Details	Compulsory 7-16, Free	Compulsory 5-16, Free
School Internet Access Percentage	100%	55% Primary 60% Secondary
Tertiary Education		
Percentage of Population expected to complete Tertiary Education –1995	67%	58%
IT Related Enrolments	8900 (1997)	5800 (1998)
Number of IT Related Graduates (1996)	2640	821

Sources: Butler & Zwimpfer, 1997; Information Technology Advisory Group, 1999; Lyytinen & Goodman, 1999; March, 1999; Ministry of Commerce, 1999; Ministry of Education, 1998; Statistics Finland, 1999a

schools have offered computing since the mid 1980s (Lyytinen & Goodman, 1999). Today, students are exposed to IT at an early age and computer literacy is part of the national curriculum. Every student has access to a computer, and every primary and secondary school has fast Web access (Information Technology Advisory Group, 1999).

New Zealand's schools are making much smaller investments in IT infrastructure than other OECD countries (Butler & Zwimpfer, 1997). In 1998, there was one computer for every 14 students in primary schools and one for every eight students in secondary schools, figures that are hardly considered adequate. Further, only 55% of primary schools and 60% of secondary schools have Web access from at least one classroom. Fortunately, there have been recent increases on IT expenditure in schools and it is expected all schools will have adequate computer and Web access within the next five years (March, 1999).

Tertiary Education

Finland has around 20 universities or other institutes of higher education. Computing education began in the higher education sector when the first chair in computing was established in 1965. By the end of 1996, IT topics were taught in 15 universities that annually graduate over 600 five-year degrees and 40 doctorates. Finland also has an extensive network of polytechnics that produce over 2,000 degrees each year in computing and engineering (Lyytinen & Goodman, 1999). Reacting to the demand for trained professionals, universities and polytechnics has dramatically expanded their computer and IT-related programs over the past few years. Finland now produces five times as many science and technology graduates as law graduates (Information Technology Advisory Group, 1999).

New Zealand currently has eight universities. Most of the universities and the 25 polytechnics offer IT-related degrees and/or diplomas (Ein-Dor et al., 1997). The number of students enrolled in IT-related courses have been increasing dramatically over the past decade. However, the 1996 graduates in these areas amounted to only 3.84% of the total number of graduates. There have also been complaints that there is a mismatch between IT graduate skills and those required by the IT industry (Ministry of Commerce, 1999).

Summary

In terms of the importance of the education system, Finland and New Zealand are very similar. Both countries have well-developed education systems, indicating that a reasonable level of education is required for IT industry success. However, the Finnish and New Zealand education systems differ in two main ways. Firstly, Finland has implemented IT and promoted IT use in schools to a greater extent than New Zealand. Secondly, Finland is producing a greater number of IT-related graduates. These two findings may even be correlated. Students that are exposed to IT in schools may be more inclined to pursue IT-related courses at tertiary level. In any case it appears that educational policies have an impact on IT industry success. We suggest that a high degree of IT competence at school and tertiary level is associated with a successful IT industry.

CONCLUSION

Given the importance of the information technology industry in today's global economy, much recent research has focused on the relative success of small countries in fostering IT industries. Those countries that can develop new information-based industries will most likely become the leaders in the new knowledge-based economy of the future. Those countries that fail to develop such industries may well find themselves at a disadvantage.

In this chapter we have examined the factors of IT industry success in small developed countries, and compared two such countries, Finland and New Zealand. Building on earlier work in this area, we used a modified version of the theoretical model suggested by Ein-Dor et al. (1997). We have suggested factors that may contribute to the differing levels of IT industry success.

This chapter has identified Finland and New Zealand as being similar in terms of country size and economic development. Despite this similarity, the two countries are experiencing differing levels of IT industry development. Finland's IT industry is far more successful than New Zealand's, particularly in terms of the hardware sector.

Though there may be other factors that have contributed to Finland's success, our research has suggested three major factors—government IT promotion, high levels of private sector R&D investment, and an education system that produces IT-literate graduates—as being important for IT industry success. These findings are consistent with those of Ein-Dor et al. (1997) and support the macroeconomic theory of Grossman and Helpman (1991). It appears that there is an optimal level of government support for IT industries in small, open economies, and that government support for the IT industry (as in Finland) is substantially better than no support at all (as in New Zealand).

This chapter has many implications for policy makers in small countries. Our findings suggest that governments need to take a proactive role in fostering the IT industry if they are to have any chance of developing information-based industries of their own. The extent of that support is still a matter for debate, but it is clear that more support is better than none at all. It is also clear that an educated workforce is a prerequisite for IT industry success, as are high levels of R&D investment.

There are two main limitations to this study. First, we openly acknowledge that the various theoretical constructs and models used to explain IT industry success are all at an early stage of development. Further research is needed into the highly complex relationships at the firm, market, policy and international levels. Second, only two small developed countries were studied in this project. This limits the generalizability of the findings. Whether our findings hold for other small developed countries requires further research.

For the future, we believe that research which examines multiple countries over an extended period of time, using researchers drawn from a variety of disciplines such as information systems, economics and public policy management, is needed. Careful, in-depth longitudinal studies will undoubtedly give us more insights into such a complex subject. At this early stage, however, exploratory reviews benchmarking the state of the IT industries in various countries—such as this chapter provides—are at least one step forward in the right direction.

ACKNOWLEDGMENTS

This chapter was previously published in the *Journal of Global Information Management*. (2001). 9(2), 4-14.

REFERENCES

Blanning, R. W., Bui, T. X. and Tan, M. (1997). National information infrastructure in Pacific Asia. *Decision Support Systems*, 21, 215-227.

Butler, G. and Zwimpfer, L. (1997). *Impact 2001: Learning with IT—The Issues*. Ministry of Commerce. Retrieved on the World Wide Web: http://www.moc.govt/pbt/infotech/ impact/ imped/html.

Central Intelligence Agency. (1999). *The World Factbook 1999*. Retrieved on the World Wide Web: http://www.odci.gov/cia/ publications/factbook/index.html.

Computerworld. (1999). *NZ Computer Industry Directory 1999*. IDG Communications.

Dedrick, J. L., Goodman, S. E. and Kraemer, K. L. (1995). Little engines that could: Computing in small energetic countries. *Communications of the ACM*, 38(5), 21-26.

Deloitte & Touche Consulting Group. (1998). Top 200 New Zealand companies. *Management Magazine,* 45, 74-87.

Dobbeck, D. and Woods, W. (1994). Mapping industrial activity. *OECD Observer*, June-July, 188, 19-23.

Edmondson, G. (1995). Oulu, Finland: In the cold and dark, high-tech heat. *Business Week*, September.

Ein-Dor, P., Myers, M. D. and Raman, K. S. (1997). Information technology in three small developed countries. *Journal of Management Information Systems*, 13(4), 61-89.

Finnfacts. (1999). *50 Largest Finnish Companies*. Retrieved on the World Wide Web: http://www.finnfacts.com/Ffeng0399/record_profits.htm.

Grossman, G. and Helpman, E. (1991). *Innovation and Growth in the Global Economy*. Cambridge, MA: MIT Press.

Helsinki Stock Exchange. (1999). *Listed Companies*. Retrieved on the World Wide Web: http://www.hex.fi/eng/listed_companies/.

IMD. (1999). *The World Competitiveness Yearbook 1999*. Lausanne.

Information Technology Advisory Group. (1999). *The Knowledge Economy*. Ministry of Commerce. Retrieved on the World Wide Web: http://www.moc.gvot.nz/pbt/infotech/itag/publications.html.

International Data Corporation. (1999). *New Zealand*.

International Telecommunication Union. (1999). *World Telecommunication Development Report-Mobile Cellular 1999*.

Kraemer, K. and Dedrick, J. (1992). *National Technology Policy and the Development of Information Industries*. Irvine, CA: Centre for Research on Information Technology and Organizations, University of California.

Kraemer, K. and Dedrick, J. (1993). Turning loose the invisible hand: New Zealand's information technology policy. *The Information Society, 9*, 365-390.

Kraemer, K. L., Gurbaxani, V. and King, J. L. (1992). Economic development, government policy and the diffusion of computing in Asia-Pacific countries. *Public Administration Review, 52*(2), 146-156.

Lyytinen, K. and Goodman, S. (1999). Finland: The unknown soldier on the IT front. *Communications of the ACM, 42*(3), 13-17.

Maenpaa, M. (1999). *Technology Provides Keys for Growth*. Virtual Finland. Retrieved on the World Wide Web: http://virtual.finland.fi/finfo/english/technology.html.

March, F. (1999). *Statistics on Information Technology in New Zealand*. Wellington, NZ: Ministry of Commerce.

Minges, M. (1999). *International Telecommunications Union*.

Ministry of Commerce. (1993). *Review of Information Technology Policy in New Zealand*. Wellington, New Zealand.

Ministry of Commerce. (1996). *Impact 2001: How Information Technology Will Change New Zealand*. Retrieved on the World Wide Web: http://www.moc.govt.nz/pbt/infotech/ impact/ impact.html.

Ministry of Commerce. (1999). *Information Technology Sector Contribution to Foresight Coordinated Through the Information Technology Advisory Group*. Retrieved on the World Wide Web: http://www.moc.govt.nz/pbt/infotech/foresight/index.html.

Ministry of Education. (1998). *Tertiary Education Statistics 1998*. Wellington, NZ: Data Management and Analysis Division.

Ministry of Finance. (1996). *Finland's Way to the Information Society-The National Strategy and its Implementation*. Retrieved on the World Wide Web: http://www.tieke.fi/tieke/tikas/indexeng.htm.

Myers, M. D. (1996). Can kiwis fly? Computing in New Zealand. *Communi-*

cations of the ACM, 39(4), 11-15.

Nygard, A. M. and Kunnas, T. (1998). *Computer Networking Hardware/ Software*. Finland: International Trade Administration.

OECD. (1996a). *Fiscal Measures to Promote R&D and Innovation*. Paris: Organization for Economic Cooperation and Development.

OECD. (1996b). *Information Infrastructure Policies in OECD Countries*. Paris: Organization for Economic Cooperation and Development.

OECD. (1996c). *Science Technology and Industry Outlook*. Paris: Organization for Economic Cooperation and Development.

OECD. (1998a). *Internet Infrastructure Indicators*. Paris.

OECD. (1998b). *Main Science and Technology Indicators*. Paris.

Perry, A. (1999). Finland: Nokia's telecommunications laboratory. *New Zealand Infotech Weekly*, December, 422.

Statistics Finland. (1999a). *On the Road to the Finnish Information Society— Summary*. Retrieved on the World Wide Web: http://www.stat.fi/tk/yr/tttietoti_en.html.

Statistics Finland. (1999b). *Statistical News*. Retrieved on the World Wide Web: http://www.stat.fi/tk/tp_tiedotteet/v99/002ttte.html.

Virtual Finland. (1999). *Factsheet Finland*. Retrieved on the World Wide Web: http://virtual.finland.fi/finfo/english/facteng.html.

World Almanac. (1997). *The World Almanac and Book of Facts 1997*. Mahwah, NJ: K-III Reference Corporation.

World Bank. (1999). *World Development Indicators 1999*. Retrieved on the World Wide Web: http://www.worldbank.org/data/wdi/.

World Economic Forum. (1998). *The Global Competitiveness Report 1998*. Geneva.

BIOGRAPHICAL SKETCH

Rebecca Watson is a Consultant in the Information Technology Management practice at PA Consulting Group, Sydney, Australia. She joined PA after completing a Bachelor's of Commerce (Honours) at the University of Auckland, New Zealand, in 1999. She majored in Information Systems, focusing on the strategy and management of information systems, e-business and operations management. Her research work has been published in the University of Auckland Business Review *and the* Journal of Global Information Management.

Michael D. Myers is Professor of Information Systems in the Department of Management Science and Information Systems at the University of Auckland, New Zealand. His research interests are in the areas of information systems development, qualitative research methods in information systems, and the social and organizational aspects of information technology. His research articles have been published in journals such as Accounting, Management and Information Technologies, Communications of the ACM, Ethics and Behavior, Information Systems Journal, Information Technology & People, Journal of Information Technology, Journal of International Information Management, Journal of Management Information Systems, Journal of Strategic Information Systems, MIS Quarterly, *and* MISQ Discovery. *He currently serves as Senior Editor of* MIS Quarterly, *Editor of the* University of Auckland Business Review, *Associate Editor of* Information Systems Research, *and Editor of the* ISWorld Section on Qualitative Research.

Identifying Supply Chain Management and E-Commerce Opportunities at PaperCo Australia

Danielle Fowler
University of Baltimore, USA

Paula M. C. Swatman
University of Koblenz-Landau, Germany

Craig Parker
Deakin University, Australia

EXECUTIVE SUMMARY

Established supply chain management techniques such as Quick Response (QR) or Customer Relationship Management (CRM) have proven the potential benefits of reorganizing an organization's processes to take advantage of the characteristics of electronic information exchange. As the Internet and other proprietary networks expand, however, organizations have the opportunity to use this enabling infrastructure to exchange other, more varied types of information than traditional electronic data interchange (EDI) messages. This is especially true of companies with global operations and interests, which lead to a more diverse set of trading activities.

This case presents the experiences of a large Australian paper products manufacturer in implementing an electronic document exchange strategy for supply chain management, including the drivers for change which spurred their actions, and describes the issues associated with trying to support existing and future requirements for document exchange across a wide variety of trading partners. The experiences of PaperCo will be relevant to organizations with diverse trading partners, especially small to medium enterprises (SMEs).

THE ISSUES ON HAND

While looking beyond established schemes for more creative and opportunistic exchanges of information with trading partners offers promise of new benefits, the proliferation of potential document exchange types and mechanisms involved have two implications: they demand a more sophisticated technological infrastructure and require a strategy for the coordinated management of the information flows themselves.

Such a strategy is especially important to organisations operating in Australia, for the following reasons:

- Australia has a very large proportion of SMEs, with a comparatively low uptake of traditional EDI (Parker, 1997), supply chain management (SCM) schemes and technologies such as the Internet, email and the Web (Pacific Access, 2000)
- While smaller businesses have had a low e-commerce adoption rate, Australia has one of the highest overall rates of IT adoption in the world, with Internet penetration ranking well ahead of comparable nations such as the UK, Taiwan, Korea, Germany and Japan. *"At 36% of the total population accessing the Internet, Australia is only behind Sweden and Canada, which are both at 43% and the U.S. at 41%. Australia is among the world leaders in accessing the Internet, whether it is measured by households or population"* (NOIE, 2000).
- The country's geographic isolation increases the population's dependence on communications and computer technologies.

The potential for less structured, less formalized (and less expensive) uses of document exchange to improve trading partner linkages in such an environment is high. Although Australia has a particular need for improved document exchange mechanisms, this situation will potentially face any large organization with a diverse set of (especially smaller) trading partners. A company in this position is faced with obvious concerns: how do we identify the types of document exchange which might be involved? How do we identify where they might occur: which parts of the organization are involved, and which processes might be improved? What is involved in the development of a document exchange strategy? How can we build a supporting infrastructure for Electronic Document Exchange (EDE)? What are the potential success factors, the functional criteria?

This chapter describes the experiences of a large Australian company which has been through this process, and presents the results of their experiences. The case description is focused around the following questions:

- *What kind of e-commerce environment do Australian or international organisations face in Australia?*
- *How might an organization find opportunities for process improvement, either internally or with regard to inter-organizational processes, by focusing on non-standardized data and document exchange, rather than by focusing on established SCM initiatives?*
- *What kind of infrastructure is required in order to enable both current and future such opportunities, given their discovery may be ad hoc, and the nature of the exchange may not be predetermined?*
- *Are multinational companies, with more complicated intra-organizational structures, or those, which engage in global commerce, with more complicated inter-organizational links, more likely to benefit from a cohesive document exchange platform and strategy?*

INTRODUCTION: EVOLVING TRADING PARTNER EXCHANGES

Organisations face constantly increasing pressure to reduce costs and to improve internal and external efficiencies so that they can remain competitive (Chatfield and Bjorn-Anderson, 1997). The inefficiencies associated with intra- and inter-organizational trade can often be attributed to delays and errors in the exchange of the business documents and information associated with the goods and services being produced and traded (Wenninger, 1999).

Many organisations have found success and a competitive advantage, or at least "kept up with the Jones's," by putting into place industry-accepted and prevalent supply chain management (SCM) approaches (initially JIT, QR; later ECR, CRP etc.—see Table 1b for definitions). They have been driven primarily through the implementation of automated and standardized document exchange (EDI). SCM has generally included business process reengineering (BPR), to streamline an organisation's value chain (Davenport and Short, 1990; Hammer and Champy, 1993; van Kirk, 1993), and the processes which connect the company and its suppliers and/or customers (Clark and Stoddard, 1996; Swatman et al., 1994).

But is simply adopting an underlying technology or choosing to seek out "industry best practices" as a blueprint for process change going to help an

Table 1: Electronic Document Exchange Tools (a) and Approaches (b)

(a)

EDE Tools	Description
Electronic Data Interchange (EDI)	The computer-based application-to-application exchange of standardised business documents such as orders and invoices, which in term reduces errors and delays due to manual rekeying of data. EDI translation software converts these business documents between proprietary used by applications and a standardised format which is exchanged (Kalakota and Whinston, 1997).
Traditional EDI	EDI exchanges occur via a proprietary value-added network (VAN).
Internet EDI	EDI exchanges occur via the Internet.
Electronic Funds Transfer (EFT)	The transfer of money from one bank account to another bank account, which can be in the same or different banks (Kalakota and Whinston, 1997; Turban et al, 2000).
Electronic Fax (e-fax)	Electronic documents (including orders) are formatted in a readable form and are received by a trading partner's facsimile machine.
Electronic Forms (e-forms)	Often take the form of web-based (as opposed to paper-based) forms which are filled out online using a web browser and submitted.

organization determine opportunities for advantage or improvement that are unique to the individual company? These are now harder to find, and more likely to involve non-standardized (non-EDI) business document exchanges.

Numerous examples of (mostly large) organisations which have implemented successful EDI projects have been reported over the last decade (see, for example DeCovny, 1998; Emmelhainz, 1992; McMichael et al., 1997; Rochester, 1989; Shaw, 1995; Swatman, 1994). Despite the ability of EDI to reduce operating costs and lead times, however, the full potential of EDI for many users has not been realized because small and medium enterprises (SMEs) in particular have resisted calls to become EDI capable (NOIE, 2000; Steel, 1996; Wenninger, 1999). EDI users have often found it necessary to operate both EDI systems as well as traditional paper-based processes to support trading partner requirements (Farhoomand and Boyer, 1994; Iacovou et al., 1995), thus reducing the effectiveness of their reengineering programs (Wenninger, 1999).

A newer trend is for organisations to undertake more complex Electronic Document Exchange[1] (EDE) with their partners than the frequently cited purchase orders and Advance Ship Notices (ASNs). For example, companies are trying to utilize their internal databases to provide trading partners with information-based products such as order tracking for customers (Baker, 1999; Rayport and Sviokola, 1995) and forecasting details for suppliers (Baljko, 1999). Catalogue and purchase-cycle solutions are commonly sought. Other organisations are streamlining their product design process through the electronic exchange of design specifications with trading partners (McCubbrey and Schuldt, 1996).

(b)

EDE Approach	Description
Business Process Reengineering (BPR)	"A management approach that focuses on the analysis and redesign of organization structures and business processes in order to achieve improvements in cost, quality, and speed" (Kalakota and Whinston, 1997, p. 27).
Just-in-Time (JIT)	Involves the delivery of a manufacturer's materials and parts at the right time, quantity and quality for use in production, so that large quantities of buffer stock is not required. This inventory management approach relies heavily on such tools as EDI for the timely exchange of orders and advance ship notices (see Turban et al, 2000; Timmers, 1999).
Quick Response (QR)	QR is the retail industry's implementation of JIT, which involves the delivery of products at the right time, quantity and quality for directly replenishing retail shelves (McMichael et al, 1997).
Efficient Consumer Response (ECR)	"... a management strategy which involves reengineering the entire grocery distribution chain to eliminate inefficiencies, excessive costs and non-value added costs for all supply chain participants" (Harris and Swatman, 1997, p. 429).
Evaluated Receipt Settlement (ERS)	"Method for initiating payment to a supplier that replaces the invoice... First the price is agreed upon by a blanket or other purchase order. Next, a material release tells the supplier the quantity to deliver. An advance ship notice confirms the quantity actually being delivered, and payment is triggered upon receipt" (ASMMA, 2001).
Continuous Replenishment Program (CRP)	"... manufacturers gain access to demand and inventory information for each downstream supply chain site [such as retailers] and make necessary modifications and forecasts for them" (Handfield and Nichols, 1999, p. 32).
Supply Chain	"... encompasses all activities associated with the flow and transformation of goods from the raw materials stage (extraction), through to the end user, as well as the associated information flows. Material and information flow both up and down the supply chain" (Handfield and Nichols, 1999, p. 2).
Supply Chain Management (SCM)	"... is the integration of these [supply chain] activities through improved supply chain relationships, to achieve a sustainable competitive advantage" (Handfield and Nichols, 1999, p. 2), which relies heavily on (some of) the tools and approaches above.

All of these EDE alternatives must be achieved in an environment where trading partners often have quite different levels of e-commerce sophistication (including EDI, Web, email or even just facsimile capability). While small businesses are increasingly making use of Internet or Web commerce (DFAT, 1999; Poon and Swatman, 1999; Pacific Access, 2000), it is still likely to be some time before all of an organisation's trading partners are Internet capable (ABS, 1999). For this reason, many large organisations are investing in e-commerce gateway solutions (Chan and Swatman, 1999; Mak and Johnston, 1999) which support this variety of EDE use.

This type of technology is also necessary to support any reengineering activities. Broadbent et al. (1999) and others suggest that organisations implementing BPR require a basic level of IT infrastructure (see also Caron et al., 1994; Dixon et al., 1994). Indeed, EDI has often been seen as a necessary precursor to the successful implementation of BPR and SCM projects, due to its ability to integrate intra- and inter- organizational systems (Swatman et al., 1994). It is our contention that EDE, which includes the electronic exchange of EDI, order tracking, product design and a plethora of other document types, will also be an important component of such an enabling IT infrastructure for BPR and SCM.

So, how might an organization approach the identification and development of such initiatives? The following sections detail the experiences of an Australian organization faced with this situation, and describe their approach and conclusions. We begin by looking at our case organization, PaperCo, and its trading environment.

CASE BACKGROUND

PaperCo is a paper products manufacturer and recycling company. It is one of Australia's largest producers of corrugated cardboard boxes, which are made from 100% recycled paper. Founded over 50 years ago, it is a privately owned, Melbourne-based international manufacturing company group employing more than 5,000 people in Australia, New Zealand and the USA, having group sales of over U.S.$1.5 billion per year. The company operates six paper recycling mills in Australia, and three in the USA (in New York, Georgia and Indiana), as well as more than 40 corrugated cardboard-box-making factories across Eastern Australia, New Zealand and the Northeast, Southeast and Midwest of the USA. The company has three main operating divisions, in addition to a head office:

1. A recycling division collects waste material (primarily paper) for recycling.
2. A paper products division takes the recycled waste and makes packaging paper.
3. A box division makes and markets corrugated box products from the recycled packaging paper. This division contains a subdivision which produces specialist printed and non-printed packaging.

We were interested in looking at PaperCo as a 'revelatory' (Yin, 1991) case for the following reasons:

• Although it has both intra- and inter-organizational international links, many of the trading partners of PaperCo are Australian SMEs. Given the poor uptake of e-commerce within SMEs, both in Australia and overseas, we were interested in whether (and how) PaperCo planned to include these trading partner relationships in their future e-commerce and SCM plans.

- In terms of size and turnover, PaperCo is a very large company by Australian standards, and has few direct competitors of any size. We were interested in evaluating the factors which were motivating PaperCo to consider document exchange and SCM innovation, to discover whether they were internal or external, local or international.

Before describing PaperCo itself, it is useful to visit the general business environment in which Australian companies operate. This environment is summarized in Table 2.

This developing, but very active, e-commerce environment has led many Australian companies—of all sizes—to investigate ways in which they can use electronic linkages to improve their internal communications as well as their trading partner relationships.

Case Analysis: Value and Supply Chain Interactions within PaperCo

Although the three divisions of PaperCo are part of the same group of companies, they are effectively separate organisations which trade heavily with one another. Much of the product produced within each company is sold down the supply chain to other divisions: the paper products divisions, for instance, sends 80% of the paper it manufactures to the box division, and sources 50% of the raw material produced by the recycling division.

Table 2: General Business Environment of Australia

Business Environment Category	Business Environment Overview
Australian population	19 million inhabitants
Australian corporate structure	96.1% of Australian businesses are small to medium (SMEs)[i]
Large and medium company use of IT[ii] as at end June 1998	100% use computers 86% use LANs/WANs 98% had Internet access (ABS, 1999)
Medium business use of eCommerce as at February 2000	89% connected to the Internet 92% using email (Pacific Access, 2000)
Small business use of eCommerce as at February 2000	60% connected to the Internet 54% using email 25% had a home page – further 18% expect to have one in 12 months (Pacific Access, 2000)

Note: i) According to unpublished Australian Bureau of Statistics figures for the 1998/99 financial year.
ii) Companies with 100 or more employees.

In exploring the possible future applications of EDE that PaperCo might implement to gain either production efficiencies or generate value-added products, it quickly became apparent that PaperCo's value and supply chains were not separate in terms of information flow, but rather quite interconnected. This causes significant interdependence, as changes to document types and flows within a single division will affect the others.

The communication between the three divisions is a combination of intra-organizational documents such as memos and general ledger transactions (the company group has centralized financial systems) and inter-organizational documents such as invoices and purchase orders. The processes, and consequently the information systems, which support these document flows are both intra- and inter-organizational systems.

PaperCo had already been using EDI to exchange a few documents with some trading partners, but the scheme had been deemed unsuccessful because of a lack of internal integration and a lack of interest from the majority of their trading partners at the time. At the commencement of our study, however, the company was facing increasingly more complex EDE requirements from its trading partners, both in terms of the quantity of documents requested and in terms of the variety of documents needed. This increase in EDE interest from trading partners had been due in large part to the advent of and interest in the Internet, and was coming primarily from local trading partners. PaperCo was also simultaneously looking at its intra- and interdivisional processes, in addition to their links with trading partners, with a view to streamlining them. A further driver for change was the result of an independent survey, which reinforced the view that PaperCo's customers were interested in value-added services which would modernize their supply chains, centered around the increased provision of documents and information. Lack of customer information, poor delivery performance, long quote cycle times and average-to-poor service performance were identified as issues in customer service management which needed to be addressed. Some of the suggested changes PaperCo was considering to address these issues were also relevant to the development of a document exchange strategy, including:
- visibility of customer information across the group,
- improved reporting of delivery performance,
- ability to assess lead-time/availability ,
- a simplified/automated quote process,
- a consistent customer management approach.

The types and quantities of the potential documents to be exchanged, both internally and externally, were diverse. The recycling division, for instance, had approximately 15,000 active customers, all with similar needs—while the box division had approximately 8,000 active customers with diverse needs and differing levels of technical sophistication. There was also a variety in the identified uses for document exchange within divisions. The box division, for instance, had begun to implement electronic exchange of box design and artwork files, greatly speeding job turnaround time, while the recycling division saw potential in providing customers with an automated facility for querying recycling pickups over the Web.

This diversity, combined with the common needs of the various parts of PaperCo, led the organization to consider ways in which it might improve its internal and inter-organizational process by means of EDE.

Identifying Opportunities for (Process) Improvement

Opportunities for EDE within PaperCo and with its trading partners were determined using two main approaches: a conventional analysis of existing document flows both internally to and externally with the organisation to identify opportunities for streamlining these flows using EDE; and a study of its trading partners to determine their EDE capabilities and requirements. The trading partner analysis was seen as especially important in the context of EDE, because the outcome of the study allowed PaperCo to ascertain the breadth of document types which would ultimately need to be supported by the EDE infrastructure. For this reason, the trading partner analysis for this EDE project was more complex than for traditional EDI projects, which tend to focus on such operational issues as (Shaw, 1995) the type of business document(s) to be implemented through EDI, and the variety of possible EDI standards used by their trading partners.

The analysis identified the following EDE types which would need to be supported:

- *facsimile*: including e-fax capabilities, but also support for clients with traditional fax;
- *email*: especially for the attachment of other electronic documents;
- *Web*: material made available by the company for customers/clients to download as they require over the Internet; the documents or information are therefore pulled by the client, where control over the access varies with the sensitivity of the data;
- *e-forms*: company-developed, structured input forms which are available to customers via the Web and which provide direct data input to PaperCo's software (which in some instances would be turned into EDI documents first); and
- *traditional EDI*.

Results from the trading partner analysis also suggested that customers were interested in value-added services which could modernize their supply chain relationships. The value-added services identified as desirable to their customers involved increased document and information provision and would therefore need to be supported by the EDE infrastructure.

An investigation of these potential value-added services, along with the internal document flow analysis, identified a number of opportunities for further EDE use within the different divisions of PaperCo:

Recycling Division

EDE use within the recycling division was limited to fax and email exchange, and the provision of recycling information on the Web. The primary users of Web material were schools, and customers seeking information. Future plans included the listing of catalogues and product information on the Web; the use of e-forms to structure requests/enquiries; EDI for exchange of invoices, purchase orders, etc., with trading partners; and increased use of email and e-fax to send/receive orders.

The most significant way in which a new EDE platform would affect the recycling division therefore would be in the ability it would give them to handle a wider range of document types, which matched its diverse customer base. It would allow them to provide commonly requested information to the community via the Internet, supporting an online marketing presence. Most additional document exchange types would centre around providing electronic (more formalized) responses to customer requests such as for scheduling and pickup information. Planned developments included the integration of recycling pickup requests submitted via the website with the scheduling system for the recycling fleet.

Historically, although they had a Web presence established, the recycling division had not been proactive in pursuing opportunistic uses for IT. This was largely because they had few people within the division with knowledge of how IT might be used; the IT requirements of the division were handled by the primary IT department of PaperCo, which is contained within the box division.

Paper Products Division

Within the paper products division, there was less variety in the uses seen for EDE, but higher volumes of a small number of document exchange types. Their document needs were based around standard business documents (invoices, orders, end-of-month statements) exchanged with a stable, smaller group of trading partners. The area of EDE expansion with the most potential was therefore EDI, as their trading partners are large organisations with relatively homogeneous needs. Again, as with the recycling division, few employees with IT expertise and no exposure to the possibilities offered had resulted in minimal awareness of the diversity of electronic exchanges possible. There was an understanding of the potential of EDI, which was expected given the EDI trial which had been conducted, but the lackluster performance of that pilot had certainly not inspired enthusiasm, and may indeed have negatively affected the generation of process change ideas within the paper products employees.

Box Division

The box division had the most potential for diverse EDE types. Their current use is focused on fax for sending/receipt of orders, but has recently widened to cover the use of fax/email in the new product design cycle (exchanging artwork and box designs with customers via email), and also on marketing opportunities similar to those of the recycling division.

The box division had a more extensive history with e-commerce via an EDI pilot, and had had the greatest number of requests for support from trading partners, with a resulting higher level of awareness of EDE opportunities. This awareness was generated primarily through individual, non-IT-related, employees who understood their areas well and saw the possibility for process improvements in their domains. These individuals had been pursuing their ideas separately, and independently from top management or the IT department. In several cases PaperCo's central administration discovered the existence of these division-based approaches only when they themselves began to look for document exchange opportunities. This knowledge and enthusiasm generated by "champions," coupled with the variety of document types they could support via EDE, meant that the box division had gone further than the other divisions in identifying opportunities to reengineer internal business processes, having established potential for redesign in the following areas:

Order Forecasting

It would aid pre-planning of factory schedules if the company knew when their customers would be placing orders. This information could be obtained by accessing customer databases directly (raising the issue of confidentiality), or by having the customer send a message with the time the order is required. Such a message could be automatically produced by the customer's internal systems, or sent manually in the form of structured email template. Structured email would reduce rekeying within the company, and could be replaced with EDI in the future for EDI-capable customers.

Order Placement

At the time of the study, orders were taken via fax, email and EDI (EDI messages were then printed and rekeyed). Further development of the company's website would enable e-forms to capture and structure this information. The installation and integration of an updated EDI gateway package would also allow EDI messages to be input directly into in-house systems, and eliminate rekeying.

Order Confirmation

A variety of confirmation types were being used within the company for different customers, from functional acknowledgments to full Purchase Order Acknowledgments (POAs), although the responses in place were not flexible. The company wanted to be able to offer the same transfer mechanisms for confirmation as for placement (i.e., fax, email and EDI) and allow trading partners to receive confirmations via whichever medium (media) they could support.

Job Tracking

PaperCo was considering altering internal job progress processes to provide customers with job-tracking facilities. Information on when an order placed would be ready was seen as highly desirable by customers engaging in tightly timetabled production schedules, and hence could provide competitive advantage. The company was weighing this advantage against the loss in flexibility in job scheduling.

Delivery

Delivery of orders was an obvious application for Evaluated Receipt Settlement for the company. ERS would allow the company to eliminate invoices; customers would read/scan the pallets of bar-coded goods as they arrived, triggering an EDI acceptance message. An EFT payment would then be generated and sent to the bank. In general, the company operates on terms of 30 days. As invoices are generated at the end of the month, the average payment return is 47 days. ERS had the potential to save the company $140,000 per month. For EFT-capable customers, the company would need to negotiate with their bank regarding detailed remittance advices associated with each electronic payment made to the bank, to ensure an adequate audit trail. For non-EFT-capable customers, invoices would still be issued (either by email or EDI).

The different opportunities identified above highlight the relatively small proportion of document exchange possibilities which are EDI-based.

While the instances where EDI could be used tended to have high-value returns (partly because a quantifiable cost-benefit analysis can be more easily done—data is more easily available), the number of uses in comparison to non-standardized electronic documents was low.

This comparatively wide variation in types and uses of documents leads naturally to the next question—that of the infrastructure required for electronic document exchange in a complex internal and inter-organizational environment.

Building a Supportive Infrastructure for EDE

What kind of infrastructure is required in order to enable current and future document exchange opportunities, given their discovery may be ad hoc, and the nature of the exchange may not be predetermined?

PaperCo approached determining the requirements for such an infrastructure in a fairly conventional way, by:

- establishing the constraints on and the success factors for the project;
- analyzing their internal and external environment to identify opportunities for streamlining business processes, using both conventional (EDI) and nontraditional electronic document exchange as focal points; and
- determining functional criteria on which to base their selection of an EDE solution.

We were primarily interested in EDE-specific aspects of these three phases, especially when compared to more traditional EDI-based projects, which would inform the development of our EDE Infrastructure Requirements Framework.

A range of factors were identified which were believed to be essential to the success and effectiveness of any EDE solution. These success factors matched the conventionally acknowledged factors suggested for any strategic IT initiative (Chan and Swatman, 1998; Frank, 1997; Galliers et al., 1995):

- without top management commitment, no major technological innovation can be successfully adopted by an organisation of any size (let al.one an organisation as large and diverse as this one);
- equally important was the issue of trading partner commitment, as successful SCM initiatives tend to have a cooperative focus;
- top-down implementation was required to ensure that all parts of the organisation install and use the chosen solution in the same way;
- since innovation cannot be effectively implemented without a common strategy, top management support for EDE innovations was required;

- because of the complexity and variety of business documents being exchanged throughout the company, implementation teams would need to include both business people and technical experts;
- a pilot study of one application within one division to identify and scope the technical and organizational difficulties which are inevitable with such a major change to operating procedures was recommended;
- once the pilot study was successfully completed, a phased implementation of the complete EDE program was recommended to ensure sufficient time and effort be given to each part of the group; and
- training was needed for both internal staff and for those trading partners using the new systems.

The SCM objectives with which the EDE infrastructure was to be aligned were also identified from the outset of the project: to reduce direct and indirect costs by means of EDE, including reductions in document processing errors; to optimize internal and external document flows using EDE to improve their SCM; to improve trading partner relationships (for example, by enhancing customer service) through their use of EDE; and to support the varying types of document exchange requirements by trading partners and by their internal supply chain, including support for future EDE requirements. Having identified these success factors and EDE objectives, PaperCo then set about determining the most pressing constraints which were to be placed on the EDE infrastructure:

1. *Outsourcing*: The EDE infrastructure developed needed to be both flexible and modular, in order to support the rapid and spontaneous development of supporting IT systems once BPR/SCM initiatives had been identified and selected. This was particularly important because any such development was likely to be outsourced.

2. *Archiving*: Many of the documents PaperCo were intending to exchange with trading partners were associated with box work designs, which would be stored on an intranet accessible to their trading partners. The number of documents involved was substantial, and the EDE infrastructure was required to cope with this, as well as possible future policy changes regarding external access to data.

3. *Staffing implications*: PaperCo wanted a solution which would require a minimum number of new staff: where possible, retraining of existing staff was preferred.

4. *Common approaches to trading partners*: PaperCo wanted a consistent approach to trading partner exchanges, to minimize staff impact and maintenance, and to facilitate the easier development of future BPR projects.

5.   *Future scalability and maintainability,*
6.   *Support for internal (as well as external) work flows,*
7.   *To avoid having separate strategies for internal and external EDE,*
8.   *Assessing the security implications of the Internet as an EDE architecture.*

During the process of interviewing staff and observing processes throughout PaperCo's operations, however, it became clear that a number of additional constraints were at least (if not more) crucial than those initially identified. The first was to consider the appropriateness of using an ISP to carry commercial-in-confidence information. PaperCo did not make significant use of the Internet, apart from email messaging and from the recycling division providing marketing material on the Web, which was accessed primarily by school children (and thus low in commercial confidentiality). Interviews with IT personnel across the company, however, made it clear that there was considerable demand to extend Internet usage beyond this level, to include confidential information. With an ISP no longer an appropriate intermediary, staff with specific Internet skills would have to be hired to build and maintain the new Internet site. The EDE solution chosen therefore needed to minimize the number of additional staff required.

These constraints (and indeed objectives) might arise in any organisation, and are not PaperCo specific. We would argue that, with the possible exception of the outsourcing constraint, all the constraints (in addition to the success factors and SCM objectives) are likely to be relevant to any company considering the development of an EDE infrastructure. This gives rise to our final question—whether multinational companies, with their more complicated intra-organizational structures, or those which engage in global commerce, with more complicated inter-organizational links, are more likely to benefit from a cohesive document-exchange platform and strategy, as has been the case with PaperCo.

Technical and Architectural Solutions

Given an understanding of both the EDE types and processes (current and new) to be supported, any architectural solution would need to be compatible with both new requirements and existing business practices. PaperCo identified a set of functional and nonfunctional criteria for the assessment of any potential infrastructure solution. The functional criteria supports:

- *VAN-based EDI:* The company has existing EDI-enabled trading partners using VANs. Any solution chosen had to provide connection and support for VAN-based EDI messages.
- *Internet EDI:* Many of PaperCo's smaller customers or suppliers have access to the Internet but would be reluctant to pay for VAN services for

relatively few transactions. Increasingly, trading partners were demanding Internet EDI.

- *Email:* Many of PaperCo's transactions with customers already occur using email, and its use will increase substantially as an exchange mechanism for binary files, particularly for off-site sales staff who will be able to cut the iterative product design development process from several visits down to one. This is a significant benefit in the case of country customers, when the sales reps can spend days driving to sites.

- *E-forms:* The company expects to use e-forms to structure standard transactions involving the website, such as queries for marketing information, and intends to expand these services to provide additional value-added information, such as updated or confirmed delivery schedules for recycling trucks.

- *WWW access:* The company required a Web server integrated with the gateway solution as a whole. The Web server enables outside access to material on the website, such as marketing material, and will also be connected to internal systems which generate the information. Placement of the Web server was expected to be on an extranet, which would enable better control over confidential information while ensuring security.

- *E-fax:* Access to fax machines across the company is sometimes variable, and many faxes are electronic documents which have been printed for transmission. E-fax would allow all employees with access to the network to send faxes from their PCs.

- *Binary file exchange:* Business documents exchanged by the company are no longer solely text-based documents. The increased exchange of documents such as artwork design or palette specifications required binary file support.

It is interesting to note that these criteria go beyond the capabilities of traditional EDI systems, so that the traditional EDI software selection criteria were insufficient for EDE. Rather, corporate gateway systems are increasingly available from e-commerce software vendors (for example, Sterling Commerce, GE Information Services, Harbinger) or are being developed in-house by organisations (Chan and Swatman, 1999; Mak and Johnston, 1999) to support the electronic exchange of the above document types.

In addition to the issue of whether each vendor could adequately support the functional EDI and Internet requirements of PaperCo, the company also had a set of business criteria to consider which are largely consistent with the criteria considered for most IT projects:

- *Adequate security*: Firewalls, extranets, levels of security and authentication, Internet and EDI.
- *Installation costs:* The company had set a relatively inflexible budget for any solution adopted, although the initial cost was of less concern than ongoing costs, and particularly possible additional staffing costs.
- *Configuration costs:* Configuration costs of different solutions varied enormously, depending on the number of document mappings required for EDI documents to be used.
- *Ongoing costs:* Included ongoing vendor and network provider charges.
- *Staffing impact:* Including not just staffing costs, but an assessment of the change in work practices of employees.
- *Service and support:* Including cost and degree of service provided, and response times.
- *Local presence:* Given its commitment to the EDE strategy, the company saw a local (within Australia) presence on the part of the vendor to be crucial.

These criteria were also largely in line with the EDE project constraints established earlier in the project, such as ensuring minimal staffing impacts.

CASE DISCUSSION

The EDE opportunities PaperCo discovered ranged from accepted supply chain initiatives such as ERS to less common ideas such as the electronic exchange of box art designs with customers or the provision of recycling pickup schedules over the Web. The support of any of these ideas separately through IT would provide the organisation with a benefit. But an EDE infrastructure which allows easy future implementation of additional process change initiatives is a knowledge-based resource of additional value to the organisation.

We have described PaperCo's approach to establishing the requirements for such a flexible EDE infrastructure, which we believe is applicable to other organizations. Many of the elements of the PaperCo approach are common to traditional IT or EDI projects, and as such are likely to be procedurally familiar to most organizations.

As far as successfully identifying EDE opportunities goes, however, the PaperCo case shows that the source of many good ideas will be individuals throughout the organization who understand and can see ways to improve specific relationships with particular trading partners. In PaperCo's case these individuals had both managerial and nonmanagerial positions, but in all cases had regular contact with their trading partners. In some instances the trading partners themselves had suggested a change. Opportunities are not likely to be uncovered with a "top-down" managerial investigation, but rather filter up/ across the organisation.

We investigated and have presented the issues which arose in the PaperCo case according to separate areas of literature and theory relevant to EDE. Initial investigations focused on strategic issues associated with SCM and BPR. The success and failure constraints derived for an EDE framework (see Table 3) show alignment with the strategic IT literature, suggesting that existing frameworks and models in this area are applicable to (at least this type of) e-commerce. The business criteria PaperCo had for an EDE approach also corresponded closely to the criteria considered for most IT projects.

At a more operational level, our investigation of the possible application areas for EDE involved process examination across all divisions of the company, and revealed functional criteria which go beyond the capabilities of traditional EDI systems, on which traditional organizational e-commerce studies have tended to focus. These findings suggest that:

- At a strategic or managerial level, the issues associated with development of an integrated EDE framework or architecture may be adequately addressed using existing BPR and SCM approaches.
- At an operational/functional level, the requirements of an EDE framework/architecture will vary from traditional EDI developments and will impact the systems development project(s) undertaken in implementing/ integrating EDE, whether they are built or bought.

The findings of the PaperCo case are summarized below, presented as a framework of requirements for an overall EDE infrastructure (in terms of both data and process).

A further implication evident in the PaperCo case study is the increasing importance of using open standards within internal systems, and the "greying" of the line between internal and external application systems. Internal systems have traditionally been isolated from external connections, and the problems of translation associated with proprietary standards (whether document, transport, etc.) has generally been addressed by the use of corporate EDI gateways. As business document exchange progresses beyond the standardized and largely static nature of EDI, however, the issue of standards is once more becoming critical. With the identification of new, ad hoc exchanges of documents with trading partners, or between internal divisions, the number and types of document and application interactions will not only increase, but will be inherently unpredictable. The interconnectivity of internal systems with those of external partners will therefore become increasingly important.

Table 3: Initial EDE Infrastructure Requirements Framework

EDE Infrastructure Requirements Determination	Further Case Notes
Verify that success factors are satisfied • top management commitment to EDE • trading partner commitment to EDE • EDE team includes technical and business people • an EDE champion • a pilot study is established to identify possible technical and organisational difficulties • a common company-wide EDE strategy is needed	PaperCo only felt ready to approach the idea of EDE-enabled process redesign when they had internal champions and when their trading partners pushed for new ways of working together. Several PaperCo employees with ideas for process redesign, however, were almost to the point of abandoning them. It is therefore important for a firm to recognise its EDE readiness and act in a timely manner. The pervasive nature of a consistent EDE approach/architecture, and the degree of inter-divisional cooperation needed to develop and sustain an effective platform also makes management support and input at the highest levels particularly critical.
Devise company-wide high-level EDE strategies • reduce direct/indirect document exchange costs • optimise internal/external document flows • improve trading partner relationships • support all necessary document types	While these high-level EDE strategies were specific to PaperCo, we believe that these strategies (and others) will most likely be applicable to all organisations trying to streamline their internal and external supply chain processes using EDE.
Determine EDE project constraints • minimise staff implications • common EDE approaches to trading partners • future scalability and maintainability • support for internal work flows • no separate internal/external EDE approaches • EDE solution must be secure (e.g. Internet)	We believe that these EDE project constraints identified by PaperCo are likely to be the same as those imposed by other companies embarking on such a project. These constraints suggest the need for EDE consistency both internally and externally to the firm, and that the EDE infrastructure must be able to support future growth.
Analyse internal and external document flows • Conduct a trading partner analysis to identify external EDE opportunities and requirements • Study internal supply chain for intra- and inter-division document exchange process changes • Identify value-added services which can be derived from information in internal databases • Encourage EDE ideas from all parts of the company rather than rely entirely on a centralised EDE strategy planning team	EDE opportunities identified by PaperCo ranged from well known initiatives such as ERS and EDI, through to emailing box designs and artwork. By focusing broadly on EDE rather than on specific approaches such as EDI or the Internet, PaperCo was able to devise a range of EDE initiatives based on externally and internally identified requirements. Indeed, the idea of artwork exchange led to the idea of developing an extranet on which their artwork catalogue could be stored, for easy retrieval or update by customers.
Evaluate EDE infrastructure business criteria (e.g.) • Adequate security • Installation, configuration and ongoing costs • Staffing impact • Service, support and local presence	While these business criteria for evaluating EDE infrastructure solutions were specific to PaperCo, they are sufficiently general that they are likely to form the basis for any organisation evaluating their own solutions.
Evaluate EDE infrastructure functional criteria (e.g.) The EDE infrastructure must support VAN-based EDI, Internet EDI, email, e-fax, e-forms, Web access, and binary file exchange.	These functional criteria devised by PaperCo were intended to support existing and future EDE needs of its internal and external supply chain. A further criteria in the selection of EDE solutions is its extendibility to handle new document types not yet envisaged. These functional criteria would constitute most of the likely EDE needs of most firms.

CONCLUSION

This chapter has presented a case study investigating the requirements associated with developing an EDE infrastructure to support SCM and BPR initiatives, particularly those connected to supporting future systems development. Such an infrastructure must facilitate the rapid and easy development of new systems which cannot necessarily be predicted far in advance, and must deal with a multitude of new document types.

There is a lack of e-commerce frameworks or models intended specifically to help organisations identify their EDE opportunities and infrastructure requirements, which might serve as an enabler of future BPR and/or SCM initiatives. Much has been written on the identification and development of such systems, but usually from a process point of view. We contend that a data-, or document-centric approach can be used to identify both new applications and a starting point for the improvement of existing processes.

We have presented a set of functional and nonfunctional criteria useful to organisations wishing to support process improvement initiatives, or simply manage a diverse document exchange portfolio, by identifying the strategic requirements of an infrastructure for the exchange of documents both internally and across their supply chain. Such an infrastructure will allow a company to support the variety of current and future strategies which may be in use across the organisation in a consistent manner (for example, cost minimization via EDI may be the focus of one division of a company, and reduced time to delivery via email exchange of product files another).

In particular, the issues identified within the PaperCo case may be helpful for large, multi-divisional organisations with disparate document exchange requirements. These are the companies most likely to benefit from EDE, and also the most likely to be in the position of exploring process redesign initiatives both internally and across their supply chain.

The analysis undertaken to develop this framework is not unique to e-Commerce but is an approach which has yielded benefits in many BPR cases (see, for example, Davenport and Short, 1990). Clearly, while there are elements of e-commerce which are unique, much of the analysis of business processes can be undertaken in a way which is already familiar to most of the companies considering the benefits of EDE. Although such an approach has been suggested by a number of authors working in the EDI environment (see, for example, Emmelhainz, 1992; Swatman, 1994; Parker, 1997), its extension to the wider field of EDE suggests that, in essence, the technologies supporting document exchange may well be less important than the process(es) implemented by the organizations or firms dealing with them.

As a single organisation, the PaperCo example can only be a revelatory case. Planned future work includes validation of our findings against other organisations in a similar position, and the extension of the EDE requirements framework to address more systems development-focused issues such as data/information architecture requirements and methodological support for developing and maintaining such an infrastructure. Government is already considering ways in which agencies can work together to maximize efficiency and minimize costs—EDE offers a potential way in which these goals may be obtained at comparatively minor cost.

ENDNOTES

1 In searching for a term which covered all types of document exchange between organisations, we failed to find one in use. We are using the term Electronic Document Exchange (EDE) to refer to the electronic exchange between trading partners of both standardized and ad hoc business documentation.

2 According to unpublished Australian Bureau of Statistics figures for the 1998/99 financial year.

3 Companies with 100 or more employees.

REFERENCES

Andel, T. (1998). EDI meets Internet. Now what? *Transportation & Distribution*, 39(6), 32-40.

ABS. (1999). *Business Use of Information Technology: Australia*. Report 8129.0, Australian Bureau of Statistics, Commonwealth of Australia, Canberra, October 5.

ASMMA (2001). "ANSI," American Supply & Machinery Manufacturers' Association (ASMMA). Retrieved March 16, 2001, on the World Wide Web: http://www.asmma.org/resources/bookstore/glossary.htm.

Baker, S. (1999). Global e-commerce, local problems. *The Journal of Business Strategy*, 20)(4), 32-38.

Baljko, J. (1999). Creativity turns EFTC supply chain to gold. *Electronic Buyers' News*, June, 48.

Broadbent, M., Weill, P. and St. Clair, D. (1999). The implication of information technology infrastructure for business process redesign. *MIS Quarterly*, 23(2), 59-182.

Caron, J., Jarvenpaa, S. and Stoddard, D. (1994). Business reengineering at CIGNA Corporation: Experiences and lessons from the first five years. *MIS Quarterly*, 18(3), 233-250.

Chan, C. and Swatman, P. M. C. (1998). EDI implementation: A broader perspective. *Bled '98–11th Bled International Conference on Electronic Commerce*, Bled, Slovenia, June 8-10, 90-108.

Chan, C. and Swatman, P. M. C. (1999). B2B e-commerce implementation: The case of BHP Steel. *ECIS '99–7th European Conference on Information Systems*, Copenhagen, Denmark, June 23-25.

Chatfield, A. and Bjorn-Andersen, N. (1997). The impact of IOS-enabled business process change on business outcomes: Transformation of the value chain of Japan Airlines. *Journal of Management Information Systems*, 14(1), 13-40.

Clark, T. and Stoddard, D. (1996). Inter-organizational business process redesign: Merging technological and process innovation. *Proceedings of the 29th Hawaiian International Conference on Systems Sciences*.

Davenport, T. and Short, J. (1990). The new industrial engineering: Information technology and business process redesign. *Sloan Management Review*, Summer, 11-27.

DeCovny, S. (1998). The electronic commerce comes of age. *The Journal of Business Strategy*, 19(6), 38-44.

Department of Foreign Affairs and Trade. (1999). *Driving Forces on the New Silk Road: The Use of Electronic Commerce by Australian Businesses*, Commonwealth of Australia, Canberra.

Dixon, J., Arnold, P., Heineke, J., Kim, J. and Mulligan, P. (1994). Business process reengineering: Improving in new strategic directions. *California Management Review*, 36(4), 93-108.

Emmelhainz, M. (1992). *EDI: A Total Management Guide, 2nd Ed.*, London, UK: International Thomson Computer Press.

Farhoomand, A. and Boyer, P. (1994). Barriers to electronic trading in Asia Pacific. *EDI Forum: The Journal of Electronic Commerce*, 7(1), 68-73.

Frank, M. (1997). The realities of Web-based electronic commerce. *Strategy & Leadership*, 25(3), 30-37.

Galliers, R. D., Swatman, P. M. C. and Swatman, P. A. (1995). Strategic information systems planning: Deriving competitive advantage from EDI. *Journal of Information Technology*, 10, (September), 149-157.

Gattorna, J. and Walters, D. (1996). *Managing the Supply Chain: A Strategic Perspective*, London, UK: Macmillan Press.

Hammer, M. and Champy, J. (1993). *Reengineering the Corporation*. New York, NY: HarperCollins.

Handfield, R. B. and Nichols, Jr, E. L. (1999). *Introduction to Supply Chain Management*, Upper Saddle River, NJ: Prentice-Hall.

Harris, J. and Swatman, P. M. C. (1997). Efficient consumer response (ECR) in Australia: The Australian grocery industry in 1996. *3rd Pacific Asia Conference on Information Systems*, Brisbane, Australia, April 1-5, 427-440.

Iacovou, C., Benbasat, I. and Dexter, A. (1995). Electronic data interchange and small organizations: Adoption and impact of technology. *MIS Quarterly*, 19(4), 465.

Kalakota, R. and Whinston, A. B. (1997). *Electronic Commerce: A Manager's Guide*, Reading, MA: Addison-Wesley.

Mak, H. and Johnston, R. (1999). Leveraging traditional EDI investment using the Internet: A case study. *32nd Hawaii International Conference on Systems Sciences*, Maui, Hawaii, January 5-8.

McCubbrey, D. and Schuldt, R. (1996). CALS: Commerce At Light Speed. *9th International Conference on EDI-IOS*, Bled, Slovenia, June 10-12, 539-546.

McMichael, H., Mackay, D. and Altmann, G. (1997). Quick response in the Australian retail supply chain. *1st Annual CollECTeR Workshop on Electronic Commerce*, Adelaide, October 3, 50-59. Retrieved on the World Wide Web: http://www.collecter.org/coll97/mackay.pdf.

NOIE. (2000). The current state of play. *National Office of the Information Economy (NOIE)*. November, Commonwealth of Australia, Canberra. Retrieved November 22, 2000, on the World Wide Web: http://www.noie.gov.au/project/information_economy/ecommerce_analysis/ie_stats/StateofPlayNov2000/index.htm.

Pacific Access. (2000). Survey of computer technology and ecommerce in Australian small and medium businesses. *Yellow Pages Small Business Index*, June. Retrieved November 22, 2000, on the World Wide Web: http://www.pacificaccess.com.au/sbi/index.html.

Parker, C. M. (1997). *Educating small and medium enterprises about electronic data interchange: Exploring the effectiveness of a business simulation approach*. Unpublished Ph.D. dissertation, Monash University, Australia.

Poon, S. and Swatman, P. M. C. (1999). An exploratory study of small business Internet commerce issues. *Information & Management*, 35(1), 9-18.

Rayport, J. and Sviokla, J. (1995). Exploiting the virtual value chain. *Harvard Business Review*, 73(6), 75-83.

Rochester, J. (1989). The strategic value of EDI. *I/S Analyzer*, 27(8).

Shaw, J. (1995). Doing business in the information age: Electronic commerce, EDI & reengineering. *Electronic Commerce Strategies*, Marietta.

Steel, K. (1996). The standardization of flexible EDI messages. In Adam, N. and Yesha, Y. (Eds.), *Electronic Commerce: Current Research Issues and Applications*, Springer, Berlin, 13-26.

Swatman, P. M. C. (1994). Business process redesign using EDI: The BHP Steel experience. *Australian Journal of Information Systems*, 1(2), 55-73.

Swatman, P. M. C., Swatman, P. A. and Fowler, D. (1994). A model of EDI integration and strategic business process reengineering. *Journal of Strategic Information Systems*, 3(1), 41-60.

Timmers, P. (1999). *Electronic Commerce: Strategies and Models for Business-to-Business Trading*, Chicester, UK: John Wiley & Sons.

Turban, E., Lee, J., King, D. and Chung, H. M. (2000). *Electronic Commerce: A Managerial Perspective*, Upper Saddle River, NJ: Prentice-Hall.

van Kirk, D. (1993). EDI could be coming to a PC near you. *Network World*, November, 30-34.

Wenninger, J. (1999). Business-to-business electronic commerce. *Current Issues in Economics and Finance*, 5(10), 1-6.

Yin, R. K. (1991). *Case Study Research: Design and Methods*, 2nd ed. London, UK: Sage Publications.

BIOGRAPHICAL SKETCHES

Danielle Fowler is an Assistant Professor within the Merrick School of Business at the University of Baltimore. She teaches MIS, MBA and webMBA programs within the school, primarily in the areas of systems development and e-commerce, which have been the focus of her research and teaching for almost a decade. She has been an active member of e-commerce and requirements engineering research groups in Australia and abroad, and has supervised research students in areas including e-commerce metrics, micropayment systems, smart cards, object-oriented design metrics and open source systems. Her primary research interests lie in the requirements engineering issues associated with inter-organisational or e-commerce systems development, and she has published and presented widely on these and other topics at academic venues around the world.

Paula M.C. Swatman is Professor of eBusiness within the Faculty of Informatics at the University of Koblenz-Landau in Germany. She is the Director of the Institute for Management, which runs the new bi-lingual degrees of Bachelor's and Master's of Science in Information Management at the University. Before moving to Germany, she was Foundation Professor and Innovation Leader in eCommerce at RMIT University, Australia. She is a recognised authority in the field of eBusiness/ eCommerce, having over 15 years' experience in this area, in which she has led a number of eCommerce research groups and conducted a number of research and consultancy projects. The main focus of both her teaching and research is the strategic implications of eBusiness/eCommerce for large and small organizations, particularly those concerning the integration of eBusiness into organisational practice and systems.

Craig Parker is a Senior Lecturer within the School of Management Information Systems at Deakin University, Australia. He is the Director of the School's Master's of Electronic Commerce program, which is offered both on campus and by distance education. He teaches in this program and in the School's eCommerce major within Deakin's Bachelor's of Commerce. Professor Parker has spent the last seven years researching business simulation approaches to teaching university students and business professionals about eCommerce. This work led to the development of a Web-based business simulation called TRECS (Teaching Realistic Electronic Commerce Solutions). He is also a Research Supervisor in such areas as virtual communities, Internet markets and e-commerce-enabled regional sustainability.

Barriers to IT Industry Success: A Socio-Technical Case Study of Bangladesh

Mahesh S. Raisinghani
University of Dallas, USA

Mahboob ur Rahman
CEO Gasmin Limited, Bangladesh

EXECUTIVE SUMMARY

Information technology (IT), which has evolved from the merger of computers, telecommunications and office automation technologies, is one of the most rapidly growing industries in the world. In Bangladesh, IT use is still in a backward stage in terms of information generation, utilization and applications. A dependable information system has not been developed for the management and operation of the government machinery and large volume of data transactions in the public/private sector organizations. There is a lack of locally and externally generated information needed for the efficient performance of the government, production, trading, service, education, scientific research and other activities of the society.

This case study of the IT scenario in Bangladesh discusses the challenges, analyzes the key issues that may be barriers to the success of its IT industry and discusses the inherent strengths which can be used as the launching pad for making Bangladesh a potential offshore source of software and data processing services. Recommended actions have been proposed under 'short term' and 'medium term,' depending on the priority and importance, and categorized by the fiscal, human resource development, infrastructure and marketing functional areas. The necessary ingredients to become a potential exporter of computer software and data processing services do not currently exist in the required quantum in Bangladesh. If Bangladesh wants to enter into this market, it needs to cross a number of hurdles within a very short time span. Concerted efforts from everybody concerned have to be put in on a war-footing basis, as this sector has the potential to generate the highest revenue for the country.

INTRODUCTION

In Bangladesh, information technology (IT) use is still in a backward stage in terms of information generation, utilization and applications. A dependable information system has not been developed for the management and operation of the government machinery and large volume of data transactions in the public/private sector organizations. There is a lack of locally and externally generated information needed for the efficient performance of the government, production, trading, service, education, scientific research and other activities of the society.

The limitation of resources, shortage of skilled manpower, inadequate research facilities and skill development programmers, lack of coordination among research organization, outmoded course curricula on science and technology education and poor social consciousness of the role of IT in nation building are the major factors contributing to this situation. Presumably, while reading this case study, the reader might be interested to know some facts such as: (1) Does Bangladesh have any historical background of computerization? (2) Was there any chronological follow up to the next generation of computerization? (3) What role did the private, autonomous and public agencies play in different periods of time for computerization? (4) What are the barriers that Bangladesh faces in attaining a sufficient pace for IT advancement? and last but not the least, (5) What is the current status of IT as a whole in the country?

This case study discusses the key issues of IT use and application in government, education and economy. Issues such as IT-based income generation, better public and private service provision, exports, fiscal and non-fiscal incentives, IT infrastructure development, human resource development, standards, protection of privacy and data security have been addressed in an objective manner.

OUTLINE/ORGANIZATION OF THIS CASE STUDY

Background

The background information presented in this section will allow the reader to get a better perspective on the information technology (IT) developments in the context of this case setting. Bangladesh emerged on the global map through a hard independence war in 1971 with Pakistan, of which it was a part. As a successor of a geographically separated country, i.e., Pakistan (then Pakistan was separated into two parts, having India in the middle with a distance of more than 1,000 miles; Bangladesh, the then East Pakistan and West Pakistan, the present Pakistan). Bangladesh, like other establishment, infrastructure and wealth, did not receive much technological wealth from its predecessor. Next we describe a chronological history of computerization in Bangladesh.

In 1964, when computers were available in North America and Europe, an IBM 1620 model was the first computer acquired by Bangladesh (then East Pakistan) by the Atomic Energy Commission, and that was the first computer in the country. Admajee Jute Mills (the largest jute mill in the world) in Bangladesh was the next acquirer of the IBM 1400 series computer in 1965, followed by the then United Bank Limited (now Agrani Bank) which procured the IBM 1900. In the early 1970s, the Bangladesh Bureau of Statistics (BBS) and the Bangladesh University of Engineering and Technology (BUET) also acquired mainframe and mini computers. In fact, until the mid-1970s computerization took place in the public sector in a very limited capacity (primarily for research, mathematical/scientific calculations and data analysis). Since then and until the mid-1980s, some international and multinational organizations started using mini and mainframe computers at the enterprise level for their research and office functions. Finally, with the evolution of ubiquitous microcomputers throughout the globe, a diffusion wave also stirred Bangladesh. This time, the nongovernment sectors, including NonGovernment Organizations (NGOs) providing support for socioeconomic development and aids in the country took the lead to accept the microprocessor-based stand-alone computers (XT and AT type) for their operations, research and office management. With the dynamic and volatile availability of microprocessors, starting from late the 1980s to present, both the government and the nongovernment sectors of the country have adopted state-of-the-art microcomputers within their organizations, albeit not abundantly.

The adoption of network technology in Bangladesh started in early 1990s when some banks and international organizations established their local area networks to connect their businesses. Very few government departments adopted the network environment. The first Internet Service Provider (ISP) of the country started its operation in 1996, prior to which some users got connected to the Internet through a dial-up ISP outside the country. At present there are more than 50 ISPs in the country, 8-10 of which are operating prominently. However, the use of the Internet has not been widespread among the computer users.

Software development industry in the country has not even progressed to meet the in-house, domestic requirements. Very few software development firms are capable of producing customized software. Formulating issues for computerization of the country needs an authorized government or autonomous section which was not present in Bangladesh until 1983, when a National Computer Committee (NCC) was formed to look into the matters, on an ad hoc basis. NCC was turned into National Computer Board (NCB) under the President's Secretariat of Bangladesh by a gazette notification of the government, but did not comprehensively define the terms and references. However, later NCC formulated some policies and guidelines for government departments to acquire computer hardware and software by a separate gazette notification on July 13, 1989. Later in 1989, through a five-day (December 17-21) long conference, NCB put some more recommendations for enhancement of the IT sector, in which formation of the Bangladesh Computer Council (BCC) and several other policy matters such as expansion of IT education and training, establishment of network, acquisition of hardware and software, and installation/implementation of computer in government and nongovernment sectors were included. In 1990, the National Parliament turned the NCB into BCC through inclusion of a new act passed to define the functions and responsibilities of its key personnel. The BCC designed an organizational structure to carry its functionalities with 37 employees as shown in Figure 1. BCC was to ensure the effective application and expansion of the use of information technology. In view of this, BCC has its impetus in formulating some policies and implementing them since its inception.

BCC proposed a budget of 250 million Taka (U.S.$5 million) in their fourth five-year plan (1990-1995) for the following five proposed IT development projects:
(1) BCC's Infrastructure Facilities
(2) National Information Technology Training Institute (NITI)
(3) National Data Network Authority (NDNA)
(4) National Software Export Bureau (NSEB)
(5) System Integrator Company Limited (SICOM)

Figure 1: BCC's Organizational Structure

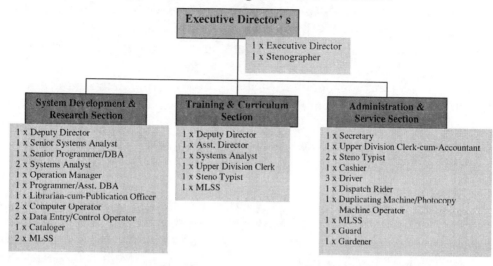

Budget allocation was made for three of these projects for financial years 1991-92 and 1992-93. BCC also proposed a new project for establishment of a "data entry industry" in the country in 1992. However, none of the BCC's projects materialized due to some political controversy.

In fact, development of the IT sector and the role of BCC seemed to fade during the period between mid-1991 to mid-1996. In late 1996, a 14-member expert committee headed by a professor of BUET and members from different business and government bodies was formed by the government to study the feasibility of software export from Bangladesh. The committee submitted a report in September 2000 to the government with some very positive recommendations to boost the IT sector in order to create a favorable environment for software industry in the country and to export software. Following the recommendations of the expert committee, the government has taken some favorable decisions from 1998 and it is being continued to date, such as the following:

(1) Declaring IT industry as a thrust sector

(2) Patronizing of IT industry by different financial sectors, especially the banks

(3) Full exemption of import tax for all IT products (e.g., hardware, software, network, peripherals and accessories)

(4) Minimizing VSAT usage charge

(5) Increase of budget allocation for IT sector

Ket Motivation for this Case Study

The key drivers for this case study can be discussed by responding to the following two questions: Why is there a need for an urgent push in IT development in Bangladesh? What benefits will it contribute to if the expected growth was achieved?

The present macroeconomic situation of the country demonstrates two broad concerns: first, fragility and low performance of several macroeconomic indicators as reflected in slow growth of investment, manufacturing output and exports; and second, underlying constraints that relate to longer run issues of efficient resource allocation, accelerated growth and sustainability of the growth process. The macroeconomic developments in the 1990s, characterized by varying and often contrasting trends in major indicators, reveal Bangladesh's continued susceptibility to economic vulnerability. Despite the government's successful mitigation of the short-term losses ensuring a much better macroeconomic performance than apprehended after the 1998 floods, several weaknesses persist in the macroeconomic balances. These are reflected in three major macro-indicators in the most recent fiscal year 2000: (i) slow growth in manufacturing output; (ii) deceleration in the rate of investment; and (iii) slow growth in exports.

In view of increasing globalization, success in export promotion of Bangladesh will largely be conditioned by its ability to integrate into the global economy. Development of software for export could provide excellent growth prospects for the Bangladesh economy. A significant mechanism of ensuring entry into global markets is through incorporation into international networks of trade and production which can be facilitated by inflow of foreign direct investment (FDI). The inflow of FDI is, however, dismally low at present.

The success of macroeconomic management is ultimately judged by its impact on the growth process and its capability to promote social goals. There seems to exist a broad consensus that a growth rate of 5-6% is not an indicator of satisfactory performance of the economy. At present, the Bangladesh economy has reached a stage that could very well yield a growth rate of 7% and above on a sustained basis, provided 'right' policies are in place.

One key question is whether the required resources are available to support such a path of growth. The gross domestic savings rate is around 18% of GDP but could go up further by 2-3% with sustained domestic savings mobilization efforts, fiscal reforms and measures to improve the performance of public sector enterprises. The current investment rate is 22% of GDP. With reforms to put the economy in a strong position to attract foreign investment and measures for foreign direct investment in infrastructure and other key areas, Bangladesh could target net foreign investment flows of about U.S.$2-3 billion per year which could contribute to raising the investment rate by about 4% of GDP. Achieving investment rates of 28-30% of GDP could support a growth rate of 7% or more, even at the current level of efficiency in capital utilization (Bangladesh Bureau of Statistics, Export Promotion Bureau, Bangladesh Bank).

The sustainability of the higher growth path, however, would require actions on several fronts, e.g., macroeconomic management that ensures stable internal and external balances, removal of infrastructural bottlenecks and ensuring socio-political stability to provide right and consistent signals to the economic agents. For this purpose, the pursuit of the reform process and ensuring credibility to the reform measures are necessary to consolidate the gains achieved during the 1990s. With the revolution in information technology in transforming the global economic structure, Bangladesh needs to contemplate a 'second generation' of reforms in the context of problems and opportunities for growth in the coming decade. These reforms should aim to acquire technological knowledge and innovations in selected areas and encourage the entrepreneurs to exploit full opportunities of the knowledge-based global economy.

PRESENT STATUS OF IT EDUCATION AND TRAINING IN BANGLADESH

The information age has raised the importance of IT skills both in the workplace and civil society. A lack of such skills on a national level will pose a barrier to Bangladesh competing effectively in the global economy. Therefore, any IT human resource development policy will encourage widespread IT literacy and an adequate supply of IT professionals to support economic goals. The computer literacy rate in Bangladesh is very low due to lack of widespread IT education within the primary and secondary schools. The country needs to produce a large number of skilled IT manpower immediately to meet this challenge.

In the context of the world scenario of human resource strength, ours is relatively poor. At the university level all national public universities and in particular new universities have started producing computer science graduates through opening computer science departments. Steps have been taken to implement computer education in the secondary and higher secondary levels. Over the last three years vigorous efforts have been made to improve the quality of IT education and research. The massive and strong participation by young people and their urge to become computer literate and develop professional competence is the most encouraging observation. Table 1 shows the number of IT graduates in 1999 from various universities in Bangladesh.

The Bangladesh Computer Council (BCC) conducts a short-duration professional efficiency enhancement training program in IT. BCC has started its standard IT training program in the Rajshahi division. Other divisional head quarters are to follow next year. The Bangladesh Institute of Communication and Information Technology (BICIT) was included in the Annual Development Program in the 1999-2000 fiscal year. This institute will impart standard IT education, training, standardization and certification of IT-related course curricula and products.

BCC has put some effort from its formation to mid-1992 and some other private firms provided some training on some programming languages and databases, but not at a great extent. Taking this opportunity, some foreign franchisers started appointing their franchisees in Bangladesh. Presently approximately 20 foreign franchisers are operating their IT training centers in Bangladesh.

To date, connection to any fiber optic cable backbone has not been possible since Bangladesh has missed the opportunity twice, first in 1992 when it passed up an almost-free-of-cost opportunity to be connected with an underground cable SEAMEU-2; and second in 1998 when it passed up a minimum-cost opportunity with SEAMEU-3. However, recently, there has been an initiation for a connection to the undersea cable network APCN-2 and a tri-party contract agreement is in process. It was expected that the job would be started in late 2001 and will be completed by end of 2002.

In formulating an IT policy in the People's Republic of Bangladesh, the government recently formed a probe committee in the Ministry of Science &

Table 1: Class of 1999 IT Graduates in the People's Republic of Bangladesh

University	No	Students per year
Public University	7	297
Private University	12	795
National University	1	600
Bangladesh Institute of Technology	4	240
Total		1932

Technology consisting of nine members appointing the Professor Jamilur Reza Choudhury as the head of the committee and the Executive Director of BCC as member secretary. It was expected that the committee would submit the final policy report to the government by Spring 2001.

ANTECEDENTS AND SCOPE OF CURRENT IT USAGE

Bangladesh had its first computer installed in the mid-1960s, which was the third-generation computer (IBM 1620), and the initial operation and calculation in computers were done through card reader and punch cards here. Chronologically minicomputers (e.g., IBM 360, 370), mainframe (e.g., IBM 4300, 4361) and microprocessor-based computers have been installed. In terms of programming languages, the third-generation programming languages FORTRAN, COBOL and RPG had been popular until the late-1980s. While the database management system used here was mainly SQL/DS, DB2 and earlier version of Oracle at that time and operating systems were CP/CMS, UNIX, AIX and PS2. However, some large organizations are continuing with some older version of UNIX, AIX and Oracle as DBMS and those third-generation programming languages. Of course some others have upgraded their systems, and also migrated to network platforms from mainframe and stand-alone environments. The current IT usage scenario at a glance can be viewed in Table 2.

With respect to the computerization of the government sector in Bangladesh, sometimes the culture of legacy and inherited practices play vital roles. In most countries, the government is the largest user of computers and related technology with the objective of enhancing public service delivery through IT. They are often expected to be model users of these technologies. Encouraging the diffusion of IT within public sector services is fundamental to supporting the social and developmental goals of the country. The application of IT within the public administration can improve efficiency, reduce the wastage of resources, enhance planning and raise the quality of services. However, the focus of most computing in the public sector has been on supporting traditional administrative and functional transactions rather than that of effective delivery of service to the public. New opportunities for improving the operations of public-sector entities and for delivering government services through electronic means must be taken. The telecom infrastructure details have been illustrated in Appendix A.

The scope of government activity varies widely, and its IT application areas are divided into the following three categories:

Table 2: The Current IT Usage Scenario in Bangladesh

Technology Usage Type		Technology Product Model
Computer	Server	IBM RS/4000 and RS/6000, Sun Enterprise, HP Netra, Compaq Proliant, Gateway ALR, Mac
	Client	Pentium (I, II & III), 486 processors, Mac
Operating Systems	Server	UNIX, AIX NT, LINUX, Mac OS, Novel Netware, OS/400
	Client	Windows 9x & NT, Mac OS (Client)
DBMS	Back Engine	Oracle, Sybase, SQL, Access, Informix
	Front Tools	Developer 2000/6i, Delphi, Power Builder, Power Designer
	Language	PL/SQL, T/SQL, Visual Basic
Programming Languages		Visual Basic, C, C++, Java, COBOL, RPG, OOP, J++
Web Design & Development Tools		FrontPage, Dream weaver, HTML Editor, XML, CSS, Java Script, ASP, Hot Dog
Drawing, Animation & Multimedia Tools		Adobe Photoshop, Visio Drawing, Illustrator, Freehand, Corel Draw, Quark Express, Macromedia Director/Flash, Auto Cad, Sound Forge
Back Office Mail Server		MS Exchange Server, Linux, IBM AX
Office Package Tools		MS Office 97/98/2000, Eudora

Clerical System

a) Office Automation Systems: Government Office
b) High-volume Transaction Systems: Bank, Insurance, Revenue Collection
c) Public Utility Services: Power, Telephone, Water/Sewer, Gas Transport, Postal, Passport/Immigration Registration, Licenses

Management Systems

a) Statistical Analytical System: Planning and Research
b) Information Resource Management: Data Resource Centre
c) Monitoring Systems: IMED
d) Computer Models for Planning Decisions: Planning Commission

Public Systems

a) Public Participation Systems: LGED
b) Sectoral Applications in Agriculture, Environment, Education, Family Planning, Health Care, Mining

CURRENT CHALLENGES

The committee formed by the Ministry of Commerce government of the People's Republic of Bangladesh has identified the major problems which are impeding the growth of software industry under four functional areas:

- Fiscal
- Human Resource Development
- Infrastructure
- Marketing

Next, each of these four functional areas is discussed in more detail.

Fiscal

a) The user base of computers is extremely low because of the high cost of computers and peripherals, due mainly to high incidence of import duty and VAT.

b) In the absence of any incentive scheme, the exporters do not feel encouraged to explore potential markets.

c) Interest rates on loans, charged by the Commercial Banks (currently between 15% to 17% p.a.) are too prohibitive for entrepreneurs to invest in the IT field.

d) The existing banking procedures are too complicated to induce exporters to bring their export remittances through banking channels.

Human Resource Development

a) Course curricula for computer-related education followed in the universities do not fully reflect the requirements of the IT industry.

b) The number of graduates in computer-related subjects produced by the universities each year is far less than the actual requirement.

c) A substantial number of such graduates leave the country for overseas employment.

d) Private IT training institutions lack the required quality of trainers.

e) Private IT training institutions do not follow any standard course curricula and examination system.

f) There is no planned scheme to increase computer literacy.

Infrastructure

a) Absence of necessary laws protecting the Intellectual Property Rights discourages prospective overseas customers from using Bangladesh as a source of supply.

b) The facility of high-speed data (both nationally and internationally) is very limited.

c)   Present cost of data communication is very high.
d)   High Speed Video Conferencing facility is not available.
e)   ISDN telecommunication line with fibre optic backbone does not exist.
f)   Resource materials on information technology, such as books, magazines, software, etc., are scanty and scattered.
g)   Whatever little hardware, software and communication resources are available cannot be found under one roof.
h)   Custom formalities for handling equipment/documents for export purpose are too time consuming to encourage export.

Marketing

a)   Bangladesh is not known to be a potential offshore source of software and data processing services.
b)   Information on prospective overseas customers is not available.
c)   Not all software in use is licensed.
d)   The use of Customized Application Software is virtually nonexistent; therefore, the domestic software market has not developed at all.
e)   It is not possible to enter into the export market without having a strong domestic market base.

ANALYSIS OF KEY ISSUES

In the past and also at present, Bangladesh has not been able to gain the optimum level of global IT status, and has always been in the bottom of the progression chart in comparison to not only the developed countries of North America and Europe, but also neighboring countries like India, Malaysia, Korea, China and Vietnam. Perhaps the IT progress in Bangladesh is hindered by other factors such as cultural, religious, historical and environmental factors. However there is no empirical evidence to support this premise. The software industry is one of the essential components of the IT industry. It is still largely dependent on human resources, and some of the developing countries are taking advantage of this opportunity. India is one of the most successful countries in developing its software industry, and in 1998-1999, it generated revenues of U.S.$6 billion (out of which export was around 40%), with an annual growth rate of 70%. In 1999-2000, India projected revenue from the software industry to be approximately U.S.$8 billion (providing round figures). In comparison, the present size of the software industry in Bangladesh is very small. Only a few firms are involved in export of software and data entry services, and the total volume of revenue generated is negligible.

India has attained this tremendous achievement through itsseamless effort starting from more than two decades back when the then country (India) policy-makers initiated, opened up and patronized the IT sector. They formulated IT policies, inspired and patronized IT entrepreneurs and professionals, including nonresident Indians. India's domestic market and infrastructure for the IT sector is well groomed, economical and easily attainable by the industry and end-users. The change-over in power and the successive governments in India have neither caused IT to step backward nor discouraged/inhibited the policies, initiation and other phenomenon related to the IT industry adopted by their predecessors. Each government that has ruled India over the past two decades tends to uphold, encourage and enhance the IT initiatives successively. Also the IT entrepreneurs and professionals have demonstrated their competence, sincerity and honesty to their profession. Being attracted by such a receptive environment to the IT industry, several international IT companies such as IBM, Microsoft, and Oracle among others have opened their branch offices in India. Table 3 highlights the domestic and export categories of the software industry in India which has helped it become a significant software producer in the world.

Table 3: Indian Software Industry

INDIAN SOFTWARE INDUSTRY [Figures in Million US Dollars]			
	95-96	96-97	99-00
	(Actual)	(Actual)	(Forecast)
I. Domestic			
Turnkey	141	189	563
Products & Packages	211	423	1,549
Consultancy	42	70	338
Data Processing	51	85	282
Others	20	23	85
Sub-Total	465	790	2,817
II. Export			
On-Site Services	394	535	1,070
Off-Shore Services	197	324	901
Off-Shore Packages	73	113	704
Others	25	42	142
Sub-Total	689	1,014	2,817
TOTAL MARKET	1,154	1,804	5,634

Source: NASSCOM, India

India has been successful in establishing itself as a major source of computer software services largely due to the timely contributions from the following groups/agencies:

Non-Resident Indians (NRIs)

Through their professional excellence and competence, NRIs created a positive impact in the international marketplace. Their patriotism and business acumen brought them back 'home,' and got them involved in this industry.

Government of India

The Department of Electronics, under the Ministry of Science and Technology, and headed by a senior Permanent Secretary, was given the required authority and freedom to create the appropriate environment for the Industry to grow. The enactment of the Software Copyright Protection Act of 1994 and its enforcement (in collaboration with NASSCOM) have been major contributors to the growth of software industry.

Educational Institutions

Universities, colleges, technical institutions, etc., both in public and private sectors, offered their total support in producing the right types of computer professionals in large numbers.

World Bank/UNDP

These international organizations helped the industry through periodic funding for strategic studies and investments. World Bank funded a number of studies on the industry, while UNDP was the main financial contributor for setting up NCST.

Venture Capital

When India was gradually establishing itself as a reliable source of supply of software services, the Government of India as well as the state governments provided a venture capital fund to augment the growth. Private venture capital was also available in abundance.

National Association of Software and Service Companies (NASSCOM)

This is the national forum representing the Computer Software and Service Industry, dedicated to the cause of protecting the interest of its members. Of all the agencies, NASSCOM probably made the most contribution in taking the industry to where it is today, through continuous dialogues and consultations with relevant government departments and other organizations.

Bangladesh has been late to patronize its IT sector; in fact it was in late 1980s when some partial IT policies were formulated by the then newly formed probe bodies (NCC, NCB & BCC). There was a lot of controversy regarding these policies and bodies due to political reasons. Later, during the period of 1991-1996, when a change of state power took place and the then new government took control, the IT sector was totally ignored with no follow up of the previous IT-related initiatives. As such, when the present government reigned in mid-1996, they had to start from scratch in finding ways and means to boost this sector. Time is needed to follow up and observe the impact of the new policies and procedures and understand its implications in the next three to five years. However, other than the opportunities that the government precipitates, the most important factor could be the attitude to accept the modern idea and advantages that the state-of-the-art technology can provide, by all levels of government, businesses and end-users. The government departments could set an example by being one of the main IT user groups, but in reality the government sector is not conducive to IT initiatives in practice. Their attitude towards accepting technology and motivation to upgrade their technical knowledge and skills seems weak. On the other hand, the supply of qualified IT professionals in the country to develop systems and applications, even to meet the limited local market demand, is seriously lagging. The IT training providers are not following any established standards to train the IT manpower, and there is a shortage of sincere and capable IT entrepreneurs in the country.

The inherent bureaucracy and traditional thinking of different departments of the government sector have become the biggest obstruction in the computerization of Bangladesh. Bangladesh is also lacking behind of proper infrastructure and logistic support to keep up with the current pace of IT development in the other parts of the world. The country is still not connected to the "information superhighway" of a fiber optic cable backbone. The PSTN (public switch telephone network) and the VSAT are regulated in monopoly by government departments and the cost is prohibitive for most IT companies.

LEVERAGING INHERENT STRENGTH FACTORS

Although not properly exploited yet, Bangladesh does have quite a few inherent strength factors which can be used as the launching pad for making this country a potential offshore source of software and data processing services. Some of these advantages are:

a) A substantial number of educated unemployed youth force, with the ability to read and write English, exists in the country. They can be trained in the required skill (particularly in data processing services) within a short time.

b) Quite a few Bangladeshi skilled professionals have been working abroad. They can be encouraged to return back to the country and/or collaborate with Bangladeshi entrepreneurs, provided the proper environment is created.

c) Universities in Bangladesh are turning out an increasing number of graduates in computer-related subjects every year, although the number is much less than the requirement.

d) A large number of Bangladeshi students are studying overseas in computer-related subjects.

e) A wide range of hardware platforms, from mainframe to PC, with a large number of Macs, are available.

f) Reasonable skills exist in the following areas :
    (i) Operating System: Windows, Windows 95, MAC/OS, Novell Netware, Windows NT, UNIX, OS/400
    (ii) Programming Language: C++, Visual Basic, Visual FoxPro, CO BOL, RPG, OOP, J++
    (iii) RDBMS: Oracle, Informix, DB/2

g) Bangladesh offers a very attractive low wage level as illustrated in Table 4.

The authors recognize that it is not possible to implement all the recommendations at once, and that all the suggested measures are not needed

Table 4: Comparative Compensation Data

	Bangladesh	India	U.S.A.
Programmers (per month)	US$ 400 to 800	US$ 1,200	US$ 4,500
Data Entry (Per 10,000 key strokes)	US$ 3 to 5	US$ 10	US$ 30 to 50

at the same time. Therefore, recommended actions need to be classified under 'short term' and 'medium term,' depending on the priority and importance. (The timeframe for short term should be a year and for medium term three to four years). In the very rapidly changing scenario of the IT Industry growth, it is very difficult to go for a longer time horizon.

KEY RECOMMENDATIONS FOR GROWTH OF BANGLADESH'S IT SECTOR

Bangladesh can learn from the Indian experience and should adopt the following measures which have helped India achieve the fast rate of growth:

1. Mobilize Non-Resident Bangladeshis (NRBs) involved in IT activities abroad. This can be done by arranging meeting/seminars in selected locations (e.g., in the Silicon Valley of California, USA) where the incentives being provided by GOB may be highlighted. These meetings should be addressed by policy makers/high officials/IT personnel representing GOB.
2. Set up Software Technology Villages with all necessary infrastructure facilities in line with STP, SEEPZ, etc.
3. Redesign course curricula of computer-related studies in the universities, colleges, etc.
4. Expand facilities in universities, colleges, etc., to produce a much larger number of computer professionals.
5. Enact appropriate laws to protect Intellectual Property Rights of computer software.
6. Take advantage of the immediate opportunities available in the following areas:
    a. Year 2000 conversion—a market of U.S.$650 billion
    b. Euro-Currency Conversion—to be required as soon as the EU agrees on single currency. Market size is still unknown.
    c. Health transcription data entry services.
7. Focus on multimedia market which is experiencing the fastest growth.
8. Participate in 'Gateway 97,' a Multimedia Exhibition to be held at the Science City of Calcutta from November 5-9, 1997.
9. Influence Microsoft Corporation of USA to involve Bangladesh in their plan to introduce Bangla as a language for Windows NT 5.0.
10. If necessary, draw from educational resources available in Calcutta to train prospective professionals.
11. Take steps for ISO 9000 Certifications, and, eventually, rating from Software Engineering Institute of Carnegie Mellon, USA. (SEI).
12. Form NASSCOM-type organization of the firms involved in software development and data processing services.

The Bangladesh government should upgrade the Bangladesh Computer Council to the level of a division, and give it the necessary authority to function as the primary facilitator to help growth of private-sector IT industry (such as National Computer Board of Singapore and Department of Electronics of India), and ask Bangladesh Computer Council to produce within 1999 at least 1,000 'trainers' capable of imparting basic computer education in the latest programming languages. It could also empower the Bangladesh Computer Council to develop a national examination and certification system for the private IT training institutions, to give certificates to those passing such examinations, and to encourage employers of both government and private sectors to give preference to such certificate holders for jobs. It could encourage software firms to form an association primarily to protect the interest of the software and data processing services sectors, in line with NASSCOM, ASOCIO, WITSA and JISA and enact appropriate laws for the protection of Intellectual Property Rights, as required under the WTO Charter.

The need for an IT policy for the development of the IT sector within the framework of overall national development is now well recognized. To this end, creation of a firm foundation for an information infrastructure in the society that meets the basic information needs for the state governance and socioeconomic activities is also recognized. The short-term and medium-term recommendations have been grouped under the following areas, the same way that the problems have been identified:

- Fiscal
- Human Resource Development
- Infrastructure
- Marketing

Given the dynamic and volatile nature of the IT industry, any long-term recommendation beyond the strong, basic commitment to IT by the Bangladesh Government for enhancing the quality of life of its people would seem futile.

Short Term

Fiscal

a)   To exempt computer hardware, software, peripherals, communication equipment, related components and spare parts thereof, from Import Duty, VAT, Infrastructure Development Surcharge, Import License Fee, Advance Income-Tax, etc.

b)   To allow Tax Holiday for the export-oriented software and data processing services industry, for 10 years (a unit will be considered export-oriented, if at least 70% of its revenue comes from export).

c)   To give a 15% price advantage (i.e., "domestic preference") to local software developers over import of the same products.

d)   To allow export of software and data processing services through Sales Contract, instead of Letters of Credit.

e)   To bring the bank interest rate on loans/advances/overdraft down to the level applicable to other export-oriented thrust sectors.

f)   To allow Special Custom Bonded Warehouse facilities for all export-oriented software houses.

g)   To create a Special Fund to be administered by the Ministry of Science & Technology for giving interest-free loans to teachers and students for purchase of computers and related equipment, through financial institutions who should be reimbursed with the interest lost.

h)   To create a Venture Capital Fund of at least Tk.10 Crore at Export Promotion Bureau for equity participation in export-oriented software and data processing services companies.

Human Resource Development

a)   To upgrade the Bangladesh Computer Council to the level of a division and to give it the necessary authority to function as the primary facilitator to help growth of private-sector IT industry (such as National Computer Board of Singapore and Department of Electronics of India).

b)   To ask the Bangladesh Computer Council to produce within 1999 at least 1,000 'trainers' capable of imparting basic computer education in the latest programming languages.

c)   To introduce 'Basic Computer Skills' as a compulsory subject for all students in all universities of the country at graduation level, starting from 1998.

d)   To introduce 'Computer Science Department' in all Polytechnics, BITs, universities and selected colleges, with at least 50 seats per class per year per institute.

e)   To ask the Bangladesh Computer Council to review 'Computer Science' course curricula currently being used in various universities after discussions with the universities, IT professionals and IT associations, keeping in view the requirements of the 21st century, and to request the concerned institutions to consider changes in their curricula in line with the suggested recommendations. Such review may be undertaken every two years.

f)   To empower the Bangladesh Computer Council to develop a national examination and certification system for the private IT training institutions, to give certificates to those passing such examinations and to encourage employers of both government and private sectors to give preference to such certificate holders for jobs.

Infrastructure

a) To enact appropriate laws for the protection of Intellectual Property Rights, as required under the WTO Charter.
b) To set up a low-cost, high-speed data and voice communication link with the USA and the UK, with a minimum speed of 2 Mbps. The private sector should be allowed to provide such service along with BTTB.
c) To set up an Internet node in the country.
d) To make Internet connectivity available at affordable rate, not exceeding Tk.0.50 (fifty paisa) per minute of use.
e) To make video conferencing facility available through VSAT.
f) To allow the private sector to set up their own satellite communication links in order to obtain competitive price advantage and greater availability of communication facilities.
g) To create separate cells at Chittagong, Dhaka, Kamalapur and Benapole Customs Houses to handle all incoming and outgoing equipment/documents/data media of the export-oriented IT industry, so as to ensure clearance of such equipment/documents within 24 hours.
h) To create a Central Resource Center at the Bangladesh Computer Council with current books, magazines, periodicals, software, manuals, etc., on IT-related subjects.
i) To encourage software firms to form an association primarily to protect the interest of the Software and Data Processing Services Sectors, in line with NASSCOM, ASOCIO, WITSA, JISA, etc.
j) To assign one Assistant Director of Export Promotion Bureau for this sector on a full-time basis.

Marketing

a) To arrange meetings/seminars in selected locations in the USA with a concentration of IT professionals of Bangladeshi origin (e.g., Silicon Valley, California, USA) to inform them about the incentives being provided by GOB and mobilize their support to help Bangladeshi entrepreneurs. These meetings should be addressed by policy makers/high officials/IT professionals representing GOB.
b) To ban use of all pirated software in all organizations, both in the public and private sectors.
c) To encourage all government, semi-government, autonomous organizations, sector corporations, banks, insurance companies, etc., to

replace the manual system of documentation and records by computerized system through the use of locally developed customized application software.

d) To send Marketing Missions to North America/EU consisting of members from IT associations and EPB, on a regular basis, with a view to publicizing Bangladesh software and data processing services capabilities as well as establishing personal contacts with the prospective customers.

e) To create a database of all major organizations/institutions engaged in outsourcing of software and data processing services, to be made jointly by EPB and Bangladesh Computer Samity (or Software Association, when it is formed), and to maintain a special homepage on the Internet.

f) To explore the possibility of obtaining business on a subcontract basis from the suppliers of software and data processing services in India, Sri Lanka and the Philippines, etc.

g) To empower the Export Promotion Bureau to ensure regular participation in all major international exhibitions/fairs for IT products and services.

h) To ask the concerned trade associations to organize international exhibitions/fairs in Bangladesh for IT products and services, in collaboration with the Export Promotion Bureau.

Medium Term

Fiscal

a) To create a Market Promotion Fund to be administered by the Export Promotion Bureau for meeting the expenses of promoting Bangladesh as a potential source of software and data processing services to the overseas markets.

b) To create a special fund for supporting industry-oriented IT research and development activities, to be administered by the Bangladesh Computer Council.

Human Resource Development

a) To introduce compulsory education in computer studies at school and college levels.

b) To strengthen the Bangladesh Computer Council and make it responsible for imparting higher level special need-based training to the IT professionals graduating from the universities. Such training programmes should be gradually extended to District Headquarters where facilities are available.

c)   To incorporate an industrial attachment program in the final year of computer science courses at the degree level.

Infrastructure

a)   To set up an Information Technology Village (ITV) at a suitable place on Tongi-Ashulia Road near Dhaka (e.g., the intersection of Tongi-Ashulia Road and the kutcha road leading to Mirpur) and to equip the same with all necessary facilities, such as high-speed communication, Special Custom Bonded Warehouse, Resource Centre (hardware, software, manuals, book), power and water supplies, telecom facilities, etc.

b)   To ask BTTB to set up ISDN/HDSN/ADSL lines all over the country, and a fibre optic backbone.

c)   To set up a Communication Hub in Bangladesh.

d)   To form a Standing Committee, with the following members, to formulate and implement policies, strategies and action plans for promotion of export of software and data processing services:
    i. Vice Chairman, Export Promotion Bureau—Convenor
    ii. President, FBCCI—Member
    iii. Executive Director, BCC—Member
    iv. Member (Customs), NBR—Member
    v. Member, BTTB—Member
    vi. A Senior Academic from the IT Field—Member
    vii. President, Bangladesh Computer Society—Member
    viii. President, Software Association—Member
    (Till a separate Association is formed, President, Bangladesh Computer Samity)

Marketing

a)   To ask the Export Promotion Bureau to set up permanent liaison offices in the USA and the UK to be manned by professional marketers of the IT field, who should perform and achieve definite quantified business objectives.

b)   To ask the Bangladesh Computer Council to create a database of Bangladesh IT Professionals working at home and abroad, in order to obtain their help when needed.

c)   To encourage IT industry members to take steps for ISO-9000 Certifications and eventually ratings from the Software Engineering Institute (SEI) of Carnegie-Mellon University, USA.

d)   To produce sufficient skilled IT professionals for export.

CONCLUSION

At the eve of the new millennium, Bangladeshis have an intuitive sense to catch up with global IT motion and notion. Though not in the past, at present a lot of decisions and steps have been taken in order to uphold and patronize the IT sector. Being a developing country, Bangladesh recognizes the need to boost its IT industry rapidly. It realizes that it has had a delayed start, but it does not want to be denied the opportunities and benefits that it could reap from IT.

A coordinated action plan involving the following agencies will produce the desired results:

Bangladesh Computer Council

This organization needs to be upgraded to a division in the Ministry and headed by a professional with the necessary authority to act as the focal point of providing required input for the industry.

Educational Institutions

Universities, BITs, colleges and polytechnics, in both public and private sectors, need to be sufficiently geared up to produce the right kind of professionals in the required numbers.

Development Partners

World Bank, ADB, UNDP, EC, OECF, JICA and other development partners should be approached to provide funds to set up necessary infrastructure, particularly for developing human resources.

Venture Capital

A Venture Capital Fund should be placed at the disposal of EPB for investment in this sector.

Non-Resident Bangladeshis (NRBs)

NRBs should be encouraged to project Bangladesh through their professional excellence, and eventually to return to the country to set up export houses.

Software Industry Association

A separate association of firms involved in the software development and data processing services industry should be established to protect the interest of this industry through maintenance of constant liaison with government and other agencies.

The necessary ingredients to become a potential exporter of computer software and data processing services do not currently exist in the required quantum in Bangladesh. If Bangladesh wants to enter into this market, it needs to cross a number of hurdles within a very short time span. Concerted efforts from everybody concerned have to be put in on a war-footing basis, as this sector has the potential to generate the highest revenue for the country.

APPENDIX A: TELECOM INFRASTRUCTURE

Telephone Status: In December 1999 Bangladesh Telegraph and Telephone Board (BTTB) had 474,322 telephone lines connected throughout Bangladesh, 61% of them being digital. BTTB is converting all of its Analogue Exchanges into Digital Exchanges, and the target is to complete all these conversion by the year 2002.

Transmission Systems in Bangladesh: Bangladesh is a country with many rivers and BTTB's long route transmission systems are mainly composed of microwave, UHF and VHF radio links. The use of optical fibre is presently limited within some city areas for interconnecting local exchange and Remote Switching Units (RSUs) in Multi Exchange Network. All administrative units are connected with their respective districts through UHF links. Most of such UHF links are digital radio system. Some of the district headquarters are connected through digital UHF links.

The major backbone transmission links in Bangladesh are presently using star formation network structure. Expansion and rehabilitation programme have been taken up for laying of 12 core optical fibre cables between Dhaka and Chittagong, which is the busiest route and still analogue.

Among the private operators, BRTA has established a microwave link between Dhaka and Sylhet. The most extensive tranmission network is being established in the private sector by Grameen Phone, which is using the Fiber Optic Cable Network of Bangladesh Railway, available along the railway route all over Bangladesh. Grameen Phone is also establishing an 140 Mbps microwave link between Khulna and Chittagong via Barisal.

International Telecommunication Facilities: Bangladesh commissioned its first Standard A Satellite Earth Station in 1975 at Betbunia to work with INTELSAT system. The international telecommunication facilities became easier and versatile after installation of the Standard B Satellite Station at Talibabad. The third Satellite Earth Station along with an International Trunk Exchange (ITX) was commissioned in 1994 at Mohakhali in Dhaka. Another direct satellite link was commissioned in between Sylhet and London since June 1995. BTTB's overseas transmission routes are mostly dependent on these four satellite earth stations working with INTELSAT satellites in IOR.

Table 1: Telecommunication Status in Bangladesh (December 1999)

Number of Telephones	602986
BTTB	474322
Private Operators	128664
Tele Density	5 per 1000
Number of Cellular Telephones	98500
Paging & Radio Trunking Subscribers	7000
Telex Subscribers	1600
Card Phones	1381
Packet Switch Subscribers	60
International Voice Circuits	2107
International Trunk Exchange	3
Total International Circuits	3936
NWD Circuits	21930
switching	61
Transmission	75
VSAT	51
Computer Penetration	1 per 7000
Internet Users	50000
Satellite Earth Station	4

These stations are characterized as follows:

Beside these satellite links there is an overseas terrestrial microwave (analogue) route, with India having 60 channels capacity to work between Dhaka and Calcutta.

Privatization of Telecommunication Services: The telecommunication sector of Bangladesh has been liberalized for private investment except for international voice communication. All other forms of communications have private-sector participation. Private telecom operators offer cellular mobile, paging and the radio truncking services as well as Internet services. Private operators are also given licenses for basic telephone services in rural areas.

Internet Services: The Internet service businesses in Bangladesh were being privately run until recently, with VAST connectivity mostly with Hong Kong and Singapore. ISPs in Bangladesh connects to the global Internet via VSAT links. BTTB also started its Internet services a year back. There are now more than 30 ISPs in Bangladesh with about 50,000 users.

REFERENCES

Bangladesh Computer Council. (2000). *Ministry of Science and Technology, Bangladesh*. Retrieved November 6, 2000, on the World Wide Web: http://www.bccbd.org/.

Heeks, R. (Ed.). (1999). *Reinventing Government In the Information Age, International Practices in IT-enabled Public Sector Reform*, Routledge Publishers.

Kirkpatrick, D. (2001). Looking for profits in poverty, *Fortune*, February 5, 174-176.

NASSCOM. (2000). *National Association of Software and Service Companies*. Retrieved November 6, 2000, on the World Wide Web: http://www.nasscom.org/.

BIOGRAPHICAL SKETCHES

Mahesh S. Raisinghani is a Faculty Member at the Graduate School of Management, University of Dallas, where he teaches MBA courses in Information Technology and e-commerce, and serves as the Director of Research for the Center for Applied Information Technology. He recently published his book, E-Commerce: Opportunities and Challenges *and has a second book,* Cases on Global E-Commerce: Theory in Action *scheduled for publication in December 2001. He has served as the Editor of the special issue of the* Journal of Electronic Commerce Research on Intelligent Agents in E-Commerce.

Mahboob ur Rahman is the CEO of Gasmin Ltd. in Bangladesh. After completing his education in Information Systems in the U.S., he returned to Bangladesh to pursue business in the IT sector. Mr. Rahman found that Bangladesh lacks skilled human resources in the IT field. He decided to work closely with the private and government bodies to establish IT training centers and educate people on modern information technology, so that Bangladeshi talent can compete in the global market. Mr. Rahman is working directly with the government to develop Bangladesh's IT infrastructure. Currently, he provides MIS solutions to government and private organizations in Bangladesh.

Index